Advanced Courses in Mathematics
CRM Barcelona

Centre de Recerca Matemàtica

Managing Editor:
Manuel Castellet

Michèle Audin
Ana Cannas da Silva
Eugene Lerman

Symplectic Geometry of
Integrable Hamiltonian Systems

Springer Basel AG

Authors' addresses:

Michèle Audin
Institut de Recherche Mathématique Avancée
Université Louis Pasteur et CNRS
7 rue René Descartes
67084 Strasbourg Cedex
France
maudin@math.u-strasbg.fr

Ana Cannas da Silva
Departamento de Matemática
Instituto Superior Técnico
Av. Rovisco Pais
1049-001 Lisboa
Portugal
acannas@math.ist.utl.pt

Eugene Lerman
Department of Mathematics
University of Illinois at Urbana-Champaign
Urbana, IL 61820
USA
lerman@math.uiuc.edu

2000 Mathematical Subject Classification 14J32, 14M25, 14R05, 52B20, 53C22, 53D10, 53D12

A CIP catalogue record for this book is available from the
Library of Congress, Washington D.C., USA

Bibliographic information published by Die Deutsche Bibliothek
Die Deutsche Bibliothek lists this publication in the Deutsche Nationalbibliografie; detailed bibliographic data
is available in the Internet at <http://dnb.ddb.de>.

ISBN 978-3-7643-2167-3   ISBN 978-3-0348-8071-8 (eBook)
DOI 10.1007/978-3-0348-8071-8

© 2003 Springer Basel AG
Originally published by Birkhäuser Verlag, Basel - Boston - Berlin 2003
Member of the BertelsmannSpringer Publishing Group
Cover design: Micha Lotrovsky, 4106 Therwil, Switzerland
Printed on acid-free paper produced from chlorine-free pulp. TCF ∞

ISBN 978-3-7643-2167-3

9 8 7 6 5 4 3 2 1   www.birkhauser-science.com

Contents

B Symplectic Toric Manifolds
Ana Cannas da Silva **85**

C   Geodesic Flows and Contact Toric Manifolds
  Eugene Lerman   **175**

Preface

This book contains an expanded version of the lectures delivered by the authors at the Euro Summer School *Symplectic Geometry of Integrable Hamiltonian Systems*. The summer school took place at the Centre de Recerca Matemàtica (Barcelona) from the 10th to the 15th of July 2001. It consisted of three main courses of 7.5 hours each and of some complementary talks dealing with integrable Hamiltonian systems.

Thus the book has three parts. The first part by Michèle Audin is devoted to *special Lagrangian submanifolds*, the second part by Ana Cannas da Silva deals with *symplectic toric manifolds*, and the last part by Eugene Lerman centers on *contact toric manifolds*. Hamiltonian systems arose, as their name suggests, from the formulation (by Hamilton) of classical mechanics: a Hamiltonian system is a dynamical system that describes the motion of a mechanical system whose total energy is conserved. This is where symplectic geometry comes from. It has since developed into an area of mathematics in its own right, but it remains a point of contact between physics and geometry. Among all the Hamiltonian systems, the *integrable* ones – those which have many conserved quantities – have special geometric properties; in particular, their solutions are very regular and quasi-periodic. Moreover, their study has been central in symplectic geometry, as it stands at a crossroads between dynamical systems, algebraic geometry, and group representation theory (to name a few areas of mathematics involved).

The quasi-periodicity of the solutions of an integrable system is a result of the fact that the system is invariant under a (semi-global) torus action. It is thus natural to investigate the symplectic manifolds that can be endowed with a (global) torus action. This leads, in a natural way, to symplectic toric manifolds (part B of this book), which are examples of extremely symmetric Hamiltonian systems. Physics makes a surprising come-back in part A: to describe Mirror Symmetry, one looks for a special kind of Lagrangian submanifolds and integrable systems, the special Lagrangians. Furthermore, integrable Hamiltonian systems on punctured cotangent bundles are a starting point for the study of contact toric manifolds (part C of this book). Along the way, tools from many different areas of mathematics are brought to bear on the questions at hand. Thus actions of Lie groups in symplectic and contact manifolds, the Delzant theorem, Morse theory, sheaves and Čech cohomology, and aspects of Calabi-Yau manifolds make an appearance in this book.

We hope that this book can serve as an introduction to symplectic and contact geometry for graduate students and that it can also be useful to research mathematicians interested in integrable systems.

Acknowledgments

This Euro Summer School was supported by the European Commission under contract number HPCF-CT-2000-00110 of the Improving Human Research Potential Programme, by the Spanish Ministry of Science and Technology under contract number PGC2000-2266-E, and by the Department of Universities, Research and Information Society of the Catalan Government under contract number ARCS01-152.

We would like to thank the Centre de Recerca Matemàtica and its director Manuel Castellet for sponsoring and hosting this Advanced Course, the CRM administrative staff Maria Julià and Consol Roca for smoothly working out innumerable details, and Carles Curràs and Eva Miranda for the mathematical organisation of the course and for making it such a pleasant experience. Thanks are also due to the participants of the course for their many comments on the material. It was indeed a very stimulating audience.

Part A

Lagrangian Submanifolds

Michèle Audin

Introduction

This text is an introduction to Lagrangian and special Lagrangian submanifolds. Special Lagrangian submanifolds were invented twenty years ago by Harvey and Lawson [18]. They have become very fashionable recently, after the work of McLean [25], leading to the beautiful speculations of Strominger, Yau and Zaslow [32] and the remarkable papers of Hitchin [19, 20] and Donaldson [11].

My aim here is mainly to present as many examples as possible. I have taken some time to explain why we know so many Lagrangian and so few special Lagrangian submanifolds and immersions. There are mainly two reasons:

- To be Lagrangian is, eventually, a *linear* property. On the other hand, the property to be special Lagrangian is, in dimension 3 and more, non linear.

- The moduli space of Lagrangian submanifolds that are close to a given one is an infinite dimensional manifold, while the corresponding moduli space of special Lagrangian submanifolds is finite dimensional.

This will be apparent in the number and nature of the examples I describe in these notes.

To prepare these lectures, in addition to the papers mentioned above, I have used standard textbooks on manifolds and vector fields as [22], on symplectic geometry as [4, 7, 24, 30] and on complex manifolds and Hodge theory as [8, 15].

I have used standard notation but, although this text pretends to be written in English, I have kept a preference for (transparent) French standards, for instance $\mathbf{P}^n(\mathbf{K})$ for the projective space of dimension n over the field $\mathbf{K}$ and ${}^tA$ for the transpose of a matrix A.

I thank Étienne Mann, Édith Socié, Thomas Vogel and Jean-Yves Welschinger for their comments and their help during the preparation of these notes. Special thanks to Mihai Damian, Alicia Jurado and Sébastien Racanière.

Chapter I

Lagrangian and special Lagrangian immersions in $\mathbf{C}^n$

In this chapter, I define Lagrangian and special Lagrangian immersions in $\mathbf{C}^n$. To begin with, I explain that $\mathbf{C}^n$ is the standard *real* vector space endowed with a non degenerate alternated bilinear form (§ I.1) and use this "symplectic structure" to define Lagrangian subspaces and immersions (§§ I.2, I.3 and I.4). Later, I use the complex structure as well, to define *special* Lagrangian immersions (§ I.5).

I.1 Symplectic form on $\mathbf{C}^n$, symplectic vector spaces

I.1.a Symplectic vector spaces

Consider the vector space $\mathbf{C}^n$ with the Hermitian form

$$\langle Z, Z' \rangle = \sum_{j=1}^{n} \bar{Z}_j Z'_j$$

(note that it is anti-linear in the first entry and linear in the second). Decompose it in real and imaginary parts:

$$\langle Z, Z' \rangle = (Z, Z') - i\omega(Z, Z').$$

The real part is the standard scalar product (Euclidean structure) of $\mathbf{C}^n = \mathbf{R}^n \times \mathbf{R}^n$,

$$(Z, Z') = \sum_{j=1}^{n} (X_j X'_j + Y_j Y'_j) = X \cdot X' + Y \cdot Y',$$

a symmetric non degenerate (real) bilinear form. The imaginary part defines a (real) bilinear form

$$\omega = \sum_{j=1}^{n}(X_j'Y_j - X_jY_j') = X' \cdot Y - X \cdot Y'$$

that is *alternated*, this meaning that $\omega(Z, Z) = 0$ for all Z. Equivalently, ω is skew-symmetric, that is,

$$\omega(Z', Z) = -\omega(Z, Z').$$

To write these formulas, I have decomposed the complex vectors of $\mathbf{C}^n$ as

$$Z = X + iY, \qquad X, Y \in \mathbf{R}^n$$

and I have used the scalar product $X \cdot Y$ of $\mathbf{R}^n$. The form ω is non degenerate too, as

$$\omega(X, Y) = 0 \text{ for all } Y \Rightarrow X = 0.$$

More generally, on a real vector space E, a *symplectic form* is a non degenerate alternated bilinear form. A vector space endowed with a symplectic form is said to be a *symplectic* vector space.

I.1.b Symplectic bases

Fix a complex unitary basis $(e_1, \ldots, e_n)$ of $\mathbf{C}^n$. Put $f_j = -ie_j$, so that

$$(e_1, \ldots, e_n, f_1, \ldots, f_n)$$

is a basis of the *real* vector space $\mathbf{C}^n$. Compute ω on the vectors of this basis:

$$\omega(e_i, e_j) = \mathrm{Im}\langle e_i, e_j \rangle = \mathrm{Im}\,\delta_{i,j} = 0,$$

also

$$\omega(f_i, f_j) = \mathrm{Im}\langle ie_i, ie_j \rangle = \mathrm{Im}\langle e_i, e_j \rangle = 0$$

and eventually

$$\omega(e_i, f_j) = \mathrm{Im}\langle e_i, -ie_j \rangle = \mathrm{Re}\langle e_i, e_j \rangle = \delta_{i,j}.$$

Inspired by these properties, we say that a basis $(e_1, \ldots, e_n, f_1, \ldots, f_n)$ of a symplectic vector space is a *symplectic basis* if

$$\omega(e_i, f_j) = \delta_{i,j} \text{ and } \omega(e_i, e_j) = \omega(f_i, f_j) = 0 \text{ for all } i \text{ and } j.$$

There are symplectic bases in all symplectic spaces, thanks to the following proposition.

Proposition I.1.1. *Let* ω *be a symplectic form on a finite dimensional vector space* E. *There exists a basis* $(e_1, \ldots, e_n, f_1, \ldots, f_n)$ *of* E *such that* $\omega(e_i, f_j) = \delta_{i,j}$ *and* $\omega(e_i, e_j) = \omega(f_i, f_j) = 0$.

Proof. As ω is non degenerate, it is not identically zero so that one can find two vectors e_1 and f_1 such that $\omega(e_1, f_1) = 1$. One then checks that the restriction of ω to the orthogonal complement (with respect to ω) of the plane $\langle e_1, f_1 \rangle$ is non degenerate. One eventually concludes by induction on the dimension — once noticed that an alternated bilinear form on a 1-dimensional vector space is zero. $\square$

In particular, the dimension of E is an even number and this is the only invariant of the isomorphism type of (E, ω). If E has dimension $2n$, then E with its symplectic form is isomorphic to $\mathbf{C}^n$ with the form ω. This result can be called a "linear Darboux theorem", in reference with the forthcoming (Darboux) theorem II.3.6.

More generally, an alternated bilinear form has a *rank*, that is the dimension of the largest subspace on which it is non degenerate, and is an even number.

Matrices

In a symplectic basis, the matrix of the symplectic form is

$$J = \begin{pmatrix} 0 & \mathrm{Id} \\ -\,\mathrm{Id} & 0 \end{pmatrix}.$$

Notice that the matrix J satisfies

$$J^2 = -\,\mathrm{Id}.$$

As the matrix of an endomorphism, this is a *complex structure*. In the symplectic basis of $\mathbf{C}^n$ associated with the canonical (complex) basis $(e_1, \ldots, e_n)$, J is nothing other that the matrix of multiplication by i.

I.1.c The symplectic form as a differential form

One can write ω as a differential form

$$\omega = \sum_{j=1}^{n} dy_j \wedge dx_j.$$

This is an *exact* differential form (the differential of a degree 1-form):

$$\omega = d\Big(\sum_{j=1}^{n} y_j dx_j\Big) = d(Y \cdot X).$$

The form $\lambda = Y \cdot dX$ is called *Liouville form* (see § II.1 below).

I.1.d   The symplectic group

This is the group of isometries of ω. A transformation g of $\mathbf{C}^n$ is *symplectic* if it satisfies

$$\omega(gZ, gZ') = \omega(Z, Z') \text{ for all } Z, Z' \in \mathbf{C}^n.$$

Call $\mathrm{Sp}(2n)$ the symplectic group of the space $\mathbf{C}^n$ of dimension $2n$. Consider all the groups $\mathrm{O}(2n)$, $\mathrm{GL}(n; \mathbf{C})$, $\mathrm{U}(n)$ and $\mathrm{Sp}(2n)$ as subgroups of $\mathrm{GL}(2n; \mathbf{R})$.

Proposition I.1.2. *The following equalities hold*

$$\mathrm{Sp}(2n) \cap \mathrm{O}(2n) = \mathrm{Sp}(2n) \cap \mathrm{GL}(n; \mathbf{C}) = \mathrm{O}(2n) \cap \mathrm{GL}(n; \mathbf{C}) = \mathrm{U}(n).$$

Proof. Let us characterize our subgroups of $\mathrm{GL}(2n; \mathbf{R})$:

(1) $g \in \mathrm{GL}(n; \mathbf{C})$ if and only if g is $\mathbf{C}$-linear, that is, if and only if

$$g(iZ) = ig(Z) \text{ for all } Z.$$

For a matrix A, this is to say that $AJ = JA$.

(2) $g \in \mathrm{Sp}(2n)$ if and only if g preserves ω, that is, if and only if $\omega(gZ, gZ') = \omega(Z, Z')$ for all Z and Z'. For a matrix A, this is

$$^t A J A = J.$$

(3) $g \in \mathrm{O}(2n)$ if and only if $(gZ, gZ') = (Z, Z')$. For a matrix A, this is $^t A A = \mathrm{Id}$.

One then checks that two of these conditions imply the third:

- (2) and (3) imply that
$$\langle gZ, gZ' \rangle = \langle Z, Z' \rangle$$
  thus that $g \in \mathrm{U}(n) \subset \mathrm{GL}(n; \mathbf{C})$.

- (3) and (1) imply that
$$\omega(gZ, gZ') = \omega(gZ, -ig(iZ')) = (gZ, g(iZ')) = (Z, iZ') = \omega(Z, Z')$$
  thus that $g \in \mathrm{Sp}(2n)$.

- in the same way, (1) and (2) imply (3).

In matrix terms, the intersection $\mathrm{Sp}(2n) \cap \mathrm{O}(2n)$ is the set of matrices

$$\begin{pmatrix} U & -V \\ V & U \end{pmatrix} \in \mathrm{GL}(n; \mathbf{C}) \subset \mathrm{GL}(2n; \mathbf{R})$$

such that

$$\begin{cases} ^t U V = {}^t V U \\ ^t U U + {}^t V V = \mathrm{Id}. \end{cases}$$

This is exactly the condition that $U + iV$ be a unitary matrix. $\square$

I.1.e Orthogonality, isotropy

Write $F^\perp$ for the Euclidean orthogonal of the real subspace F of $\mathbf{C}^n$ and F° for its symplectic (that is, with respect to ω) orthogonal. As ω is non degenerate, one has

$$(F^\circ)^\circ = F \qquad \text{and} \qquad \dim F + \dim F^\circ = 2n = \dim_{\mathbf{R}} \mathbf{C}^n.$$

Notice however that a subspace and its orthogonal may have a non trivial intersection. The restriction of the non degenerate form ω to a subspace is not always a non degenerate form, in contradiction with what happens in the Euclidean case (which is due to the positivity of the scalar product). In other words, all the subspaces of a symplectic space do not have the same behaviour with respect to the symplectic form. See Exercises I.6 and I.7.

One says that a subspace F is *isotropic* if $F \subset F^\circ$, *co-isotropic* if $F \supset F^\circ$. For instance, a (real) line is always isotropic, as it lies in its orthogonal which is a (real, co-isotropic) hyperplane. Notice that F is isotropic if and only if F° is co-isotropic. Notice also that the dimension of an isotropic subspace is at most equal to n, half the dimension of $\mathbf{C}^n$.

I.2 Lagrangian subspaces

I.2.a Definition of Lagrangian subspaces

The isotropic subspaces of maximal dimension n are *Lagrangian*. For instance, $\mathbf{R}^n \subset \mathbf{C}^n$ is a Lagrangian subspace. More generally, a subspace generated by "one half" of a symplectic basis is Lagrangian. Conversely, if F is an isotropic subspace of dimension $k \le n$, it is possible to complete any basis $(e_1, \ldots, e_k)$ of F in a symplectic basis and thus to obtain Lagrangian subspaces containing F.

Let us use now the complex multiplication in $\mathbf{C}^n$ to state:

Lemma I.2.1. *A real subspace P of $\mathbf{C}^n$ is Lagrangian if and only if $P^\perp = iP$.*

Proof. This is a straightforward computation:

$$\omega(Z, Z') = 0 \Leftrightarrow \operatorname{Im}\langle Z, Z' \rangle = 0$$
$$\Leftrightarrow \operatorname{Re}\langle Z, iZ' \rangle = 0$$
$$\Leftrightarrow (Z, iZ') = 0.$$

$\square$

Lemma I.2.2. *Let P be a Lagrangian subspace of $\mathbf{C}^n$ and let $(x_1, \ldots, x_n)$ be an orthonormal basis of this real subspace. Then $(x_1, \ldots, x_n)$ is a complex unitary basis of $\mathbf{C}^n$. Conversely, if $(x_1, \ldots, x_n)$ is a unitary basis of $\mathbf{C}^n$, the real subspace it spans is Lagrangian.*

Proof. If $(x_1, \ldots, x_n)$ is an orthonormal basis of the Lagrangian P, the previous lemma says that the basis $(x_1, \ldots, x_n, ix_1, \ldots, ix_n)$ is an orthonormal basis of the real space $\mathbf{C}^n$, thus that $(x_1, \ldots, x_n)$ is a complex basis of $\mathbf{C}^n$. Moreover, one has

$$\langle x_i, x_j \rangle = (x_i, x_j) - i\omega(x_i, x_j) = \delta_{i,j} - 0,$$

thus this is a unitary basis. The converse is even more obvious.   □

I.2.b   The symplectic reduction

This is a simple but useful operation, essentially contained in the next lemma.

Lemma I.2.3. *Let P be a Lagrangian subspace and F be a co-isotropic subspace of* $\mathbf{C}^n$, *such that*

$$P + F = \mathbf{C}^n.$$

Then the restriction of the projection

$$P \cap F \subset F \longrightarrow F/F^\circ$$

is injective, the space F/F° is symplectic and the image of $P \cap F$ is a Lagrangian subspace.

Proof. The symplectic form of $\mathbf{C}^n$ clearly induces a non degenerate form on F/F°, as F° is the kernel of the restriction of ω to F. The kernel of the composition

$$P \cap F \subset F \longrightarrow F/F^\circ$$

is

$$
\begin{aligned}
P \cap F \cap F^\circ &= P \cap F^\circ && F \text{ being co-isotropic, } F \supset F^\circ \\
&= (P^\circ + F)^\circ, && \text{since } (A+B)^\circ = A^\circ \cap B^\circ, \\
&= (P + F)^\circ && \text{as } P \text{ is Lagrangian, } P = P^\circ \\
&= (\mathbf{C}^n)^\circ && \text{because } P + F = \mathbf{C}^n \\
&= 0 && \text{as } \omega \text{ is non degenerate.}
\end{aligned}
$$

The map is thus injective. Eventually $P \cap F$ is isotropic and has dimension

$$\dim P \cap F = \dim P + \dim F - \dim(P + F) = \dim F - n,$$

half the dimension of the symplectic space F/F°, that is

$$\dim F/F^\circ = \dim F - (2n - \dim F) = 2(\dim F - n).$$

□

See more generally Exercise I.9.

I.3 The Lagrangian Grassmannian

We consider now the set Λ_n of all Lagrangian subspaces of $\mathbf{C}^n$.

I.3.a The Grassmannian Λ_n as a homogeneous space

Look again at lemma I.2.2. If P_1 and P_2 are two Lagrangian subspaces of $\mathbf{C}^n$, choose an orthonormal basis for each. We thus have two unitary bases of $\mathbf{C}^n$. There exists a unitary transformation (an element of the unitary group $\mathrm{U}(n)$) that maps the basis of P_1 on that of P_2... and thus *a fortiori* the Lagrangian P_1 on the Lagrangian P_2.

In other words, the group $\mathrm{U}(n)$ *acts transitively* on the set of Lagrangian subspaces of $\mathbf{C}^n$. The stabilizer of the Lagrangian $\mathbf{R}^n$ is the group $\mathrm{O}(n)$ of orthonormal basis changes in $\mathbf{R}^n$. We have defined this way a bijection

$$\mathrm{U}(n)/\mathrm{O}(n) \longrightarrow \Lambda_n$$

with the help of which we identify the two sets. Notice that this provides Λ_n with a topology, namely that of $\mathrm{U}(n)/\mathrm{O}(n)$, the quotient topology of the topology of the matrix group $\mathrm{U}(n)$.

Example I.3.1. As all lines are isotropic, the space Λ_1 is the space of real lines in $\mathbf{C} = \mathbf{R}^2$, namely the projective space $\mathbf{P}^1(\mathbf{R})$. The unitary group $\mathrm{U}(1)$ is a circle and the orthogonal group $\mathrm{O}(1)$ is the group with two elements $\{\pm 1\}$.

As the unitary group $\mathrm{U}(n)$ is compact (being closed and bounded in the space of matrices) and path-connected (exercise), the space Λ_n is a compact path-connected topological space.

I.3.b The manifold Λ_n

Let us firstly describe a neighbourhood of $P \in \Lambda_n$ in Λ_n. Put

$$U_P = \{Q \in \Lambda_n \mid Q \cap (iP) = 0\}.$$

This is an open subset: using a unitary matrix, one can assume that $P = \mathbf{R}^n$, but then $U_{\mathbf{R}^n}$ is the image in Λ_n of the (saturated) open subset of $\mathrm{U}(n)$ consisting of all the unitary bases the real parts of whose vectors form a basis of $\mathbf{R}^n$. This is, clearly, a neighbourhood of P.

Lemma I.3.2. *The open set U_P is homeomorphic to the real vector space of all symmetric endomorphisms of P.*

Proof. The subspaces Q that intersect iP only at 0 are the graphs of the linear maps $\varphi : P \to iP$. It is more convenient to call $i\varphi$ the linear map, so that φ is a linear map from P to itself. Write now that Q is Lagrangian, namely that

$$\forall\, x, y \in P, \quad \omega(x + i\varphi(x), y + i\varphi(y)) = 0.$$

We have

$$\begin{aligned}
\omega(x + i\varphi(x), y + i\varphi(y)) &= -\operatorname{Im}\langle x + i\varphi(x), y + i\varphi(y)\rangle \\
&= \omega(x, y) + \omega(\varphi(x), \varphi(y)) + (\varphi(x), y) - (x, \varphi(y)) \\
&= (\varphi(x), y) - (x, \varphi(y)),
\end{aligned}$$

P being Lagrangian. The subspace Q is Lagrangian if and only if the last expression vanishes for all x and y in P, namely if and only if φ is symmetric[1]. We have thus defined a bijection that maps 0 to P

$$\begin{aligned}
\operatorname{End}\operatorname{Sym}(P) &\longrightarrow U_P \\
\varphi &\longmapsto \text{graph of } i\varphi
\end{aligned}$$

and is clearly a homeomorphism.   □

Remark I.3.3. Consider for instance the "vertical" Lagrangian $i\mathbf{R}^n \subset \mathbf{C}^n$. We see that Λ_n is a disjoint union

$$\Lambda_n = \Lambda_n^0 \cup \Sigma_n$$

where Σ_n is the set of all Lagrangians that are not transversal to $i\mathbf{R}^n$ and Λ_n^0 is identified with the space of $n \times n$ real symmetric matrices.

We intend to prove now that the open sets U_P define the structure of a manifold on Λ_n. Notice firstly that any n-dimensional subspace Q of $\mathbf{R}^n \times \mathbf{R}^n$ may be represented by a rank-n matrix

$$Z = \begin{pmatrix} X \\ Y \end{pmatrix}, \text{ with } 2n \text{ lines and } n \text{ columns,}$$

the column vectors of which form a basis of Q. Two matrices Z and Z' describe the same subspace if and only if there exists an $n \times n$ invertible matrix $g \in \operatorname{GL}(n; \mathbf{R})$, such that $Zg = Z'$.

Lemma I.3.4. *The subspace Q is Lagrangian if and only if the two matrices X and Y are such that*

$${}^t XY = {}^t YX.$$

Proof. Let $u, u' \in \mathbf{R}^n$ and let z, z' be the corresponding vectors in Q:

$$z = \begin{pmatrix} X \\ Y \end{pmatrix} u, \quad z' = \begin{pmatrix} X \\ Y \end{pmatrix} u'.$$

Note that Xu, Yu, Xu' and Yu' are vectors of $\mathbf{R}^n$. We compute:

$$\begin{aligned}
\omega(z, z') &= \omega((Xu, Yu), (Xu', Yu')) \\
&= (Xu) \cdot (Yu') - (Yu) \cdot (Xu') \text{ (scalar product in } \mathbf{R}^n) \\
&= {}^t u \, {}^t XY u' - {}^t u \, {}^t Y X u' \text{ (as } U \cdot V = {}^t UV) \\
&= {}^t u \left({}^t XY - {}^t YX \right) u'.
\end{aligned}$$

□

---

[1] See also Exercise I.8.

Remark I.3.5. If Q is the graph of a linear map $\mathbf{R}^n \to i\mathbf{R}^n$, it can be represented by a matrix $Z = \begin{pmatrix} \mathrm{Id} \\ A \end{pmatrix}$. The relation in lemma I.3.4 simply expresses the fact that the matrix A is symmetric.

Consider more generally a subset J of $\{1, \ldots, n\}$ and the Lagrangian subspace P_J of $\mathbf{R}^n \times \mathbf{R}^n$ spanned by $\{(e_j)_{j \in J}, (ie_j)_{j \notin J}\}$. Denote U_{P_J} by U_J (for simplicity). Any element of U_J is described by a unique matrix Z such that, if we extract from Z the matrix containing the lines j (for $j \in J$) and $j + n$ (for $j \notin J$), we get the identity matrix. The 2^n open sets U_J clearly cover Λ_n. Moreover, as we have said it, each of them can be identified with the subspace $\mathrm{Sym}(n; \mathbf{R})$ of $n \times n$ symmetric matrices. The U_J's, with their identification with $\mathrm{Sym}(n; \mathbf{R})$ are coordinate charts. Change of coordinates are given by

$$
\mathrm{Sym}(n; \mathbf{R}) \xrightarrow{\varphi_J^{-1}} U_J \cap U_{J'} \xrightarrow{\varphi_{J'}} \mathrm{Sym}(n; \mathbf{R})
$$
$$
A \longmapsto Z_J(A) = Z_{J'}(B) \longmapsto B
$$

where $Z_J(A)$ is the matrix obtained from $\begin{pmatrix} \mathrm{Id} \\ A \end{pmatrix}$ by mapping the first n lines on the lines j (for $j \in J$) and $j + n$ (for $j \notin J$). The matrix $Z_{J'}(B)$ is obtained by multiplying $Z_J(A)$ by the inverse matrix of the (invertible!) matrix of the lines corresponding to J' in $Z_J(A)$. The coordinate change $A \mapsto B$ is clearly smooth (it is actually rational, thus analytic).

Proposition I.3.6. *The Grassmannian* Λ_n *is a compact and connected manifold of dimension* $\dfrac{n(n+1)}{2}$. $\qquad\square$

I.3.c The tautological vector bundle

Consider the space
$$
E_n = \{(P, x) \in \Lambda_n \times \mathbf{C}^n \mid x \in P\}.
$$

Together with its projection on Λ_n, this is a rank-n vector bundle over Λ_n. The fiber of E_n at $P \in \Lambda_n$ is the Lagrangian subspace P itself, a reason why this bundle is qualified as "tautological".

The property expressed in Lemma I.2.1, namely $P^\perp = iP$, is translated, in terms of the bundle E_n, in the fact that $E_n \otimes_{\mathbf{R}} \mathbf{C}$, the complexified bundle, is trivial (has a canonical trivialization). The (global) trivialization is the isomorphism of complex vector bundles

$$
E_n \otimes_{\mathbf{R}} \mathbf{C} \longrightarrow \Lambda_n \times \mathbf{C}^n
$$
$$
(P, x \otimes (a + ib)) \longmapsto (P, (a + ib)x).
$$

I.3.d   The tangent bundle to Λ_n

The canonical identification of the open subset U_P with the space of symmetric endomorphisms of P allows to identify the tangent bundle of Λ_n with the bundle $\operatorname{End}\operatorname{Sym}(E_n)$. It is also possible to describe this bundle from the tangent bundle of $U(n)$. The group $U(n)$ is described as a submanifold of the space of all complex matrices by the equation $^t\bar{A}A = \operatorname{Id}$, so that we have

$$T_A\,U(n) = \left\{ X \in GL(n;\mathbf{C}) \mid {}^t\bar{A}X + {}^t\bar{X}A = 0 \right\}.$$

Call $\mathfrak{u}(n)$ the vector space $T_{\operatorname{Id}}U(n)$ of skew-Hermitian matrices. There is an isomorphism

$$
\begin{aligned}
T_A\,U(n) &\longrightarrow \mathfrak{u}(n)\\
X &\longmapsto {}^t\bar{A}X
\end{aligned}
$$

identifying the tangent bundle $T\,U(n)$ with the trivial bundle $U(n) \times \mathfrak{u}(n)$ — as any Lie group, $U(n)$ is parallelizable. Consider the Lagrangian $\mathbf{R}^n$, image in Λ_n of the identity matrix Id. One can write

$$T_{\mathbf{R}^n}(U(n)/\,O(n)) = \mathfrak{u}(n)/\mathfrak{o}(n),$$

this is the quotient of the vector space of anti-Hermitian matrices by that of skew-symmetric real matrices. We thus identify

$$T_{\mathbf{R}^n}\Lambda_n = i\operatorname{Sym}(n;\mathbf{R}),$$

as the real part of a skew-Hermitian matrix is skew-symmetric and its imaginary part is symmetric.

Let P be any Lagrangian subspace. Choose a unitary matrix A such that $P = A \cdot \mathbf{R}^n$. As we have identified the quotient $\mathfrak{u}(n)/\mathfrak{o}(n)$ with the subspace $i\operatorname{Sym}(n;\mathbf{R})$ of $\mathfrak{u}(n)$, we identify the quotient $T_{[A]}\Lambda_n$ with a subspace of $T_A\,U(n)$:

$$
\begin{array}{ccc}
i\operatorname{Sym}(n;\mathbf{R}) & \xrightarrow{\;\;X \mapsto A \cdot X\;\;} & T_{[A]}\Lambda_n\\[4pt]
\Big\downarrow & & \Big\uparrow\\[4pt]
\mathfrak{u}(n) & \longrightarrow & T_A\,U(n).
\end{array}
$$

We derive an isomorphism

$$
\begin{aligned}
i\operatorname{Sym}(n;\mathbf{R}) &\longrightarrow T_P\Lambda_n\\
X &\longmapsto A \cdot X.
\end{aligned}
$$

Remark I.3.7. This isomorphism depends on the choice of A, this is why it does not follow that Λ_n is parallelizable (it is actually not, as soon as $n \geq 2$).

I.3.e   The case of oriented Lagrangian subspaces

One can also consider the space $\widetilde{\Lambda}_n$ of *oriented* Lagrangian subspaces. Replacing "orthonormal basis" by "positive orthonormal basis" in what precedes, we get an identification of $\widetilde{\Lambda}_n$ with $\mathrm{U}(n)/\mathrm{SO}(n)$.

I.3.f   The determinant and the Maslov class

The "determinant" mapping

$$\det : \mathrm{U}(n) \longrightarrow S^1$$

descends to the quotient by $\mathrm{SO}(n)$ and, in the same way, its square

$$\det^2 : \mathrm{U}(n) \longrightarrow S^1$$

to the quotient by $\mathrm{O}(n)$. This allows to compute the fundamental groups of Λ_n and $\widetilde{\Lambda}_n$.

Proposition I.3.8. *The fundamental group of Λ_n (resp. $\widetilde{\Lambda}_n$) is isomorphic to $\mathbf{Z}$. The covering $\widetilde{\Lambda}_n \to \Lambda_n$ shows $\pi_1(\widetilde{\Lambda}_n)$ as an index-2 subgroup in $\pi_1(\Lambda_n)$.*

Proof. Recall first that the group $\mathrm{SU}(n)$ is simply connected. This can be proved by induction on n: $\mathrm{SU}(1)$ is a point and $\mathrm{SU}(n+1)$ acts transitively on the unit sphere S^{2n+1} of $\mathbf{C}^{n+1}$ with stabilizer $\mathrm{SU}(n)$, so that the exact sequence

$$\pi_1 \mathrm{SU}(n) \longrightarrow \pi_1 \mathrm{SU}(n+1) \longrightarrow \pi_1 S^{2n+1}$$

gives the result. As the determinant mapping

$$\det : \mathrm{U}(n) \longrightarrow S^1$$

is a fibration with fiber $\mathrm{SU}(n)$, it induces an isomorphism

$$\det_\star : \pi_1 \mathrm{U}(n) \longrightarrow \pi_1(S^1).$$

The fiber of the determinant mapping $\widetilde{\Lambda}_n \longrightarrow S^1$ is $\mathrm{SU}(n)/\mathrm{SO}(n)$, which is simply connected, thus

$$\det_\star : \pi_1 \widetilde{\Lambda}_n \longrightarrow \pi_1 S^1$$

is an isomorphism. What is left to prove is a consequence of the fact that the diagram

$$
\begin{array}{ccc}
\widetilde{\Lambda}_n & \longrightarrow & \Lambda_n \\
{\scriptstyle \det} \downarrow & & \downarrow {\scriptstyle \det^2} \\
S^1 & \xrightarrow{\ z \mapsto z^2\ } & S^1
\end{array}
$$

is commutative. $\qquad\square$

"The" generator of $\pi_1\Lambda_n$ is called the *Maslov class*. One also calls "Maslov class" the cohomology class

$$\mu \in H^1(\Lambda_n; \mathbf{Z})$$

that it defines by duality. Using the notation of Remark I.3.3, it can be shown that μ is the dual class to the integral homology class represented by Σ_n (see [1, 12]).

I.4   Lagrangian submanifolds in $\mathbf{C}^n$

We are going now to globalize the notion of Lagrangian subspace, considering submanifolds of $\mathbf{C}^n$ whose tangent space at any point is Lagrangian. We will not really need actual submanifolds, but maps

$$f : V \longrightarrow \mathbf{C}^n$$

from some n-dimensional manifold to $\mathbf{C}^n$, the tangent mapping of which

$$T_x f : T_x V \longrightarrow \mathbf{C}^n$$

is an injection for any point x of V, with image a Lagrangian subspace. It is then said that f is a *Lagrangian immersion*.

For instance, any immersion of a curve (real manifold of dimension 1) in $\mathbf{C}$ is a Lagrangian immersion. Any product of Lagrangian immersions is a Lagrangian immersion (into the product target space), we thus obtain Lagrangian immersions of tori (products of circles). Our next aim is to describe examples of Lagrangian submanifolds and immersions in $\mathbf{C}^n$ and to give a necessary (and sufficient) condition for a given manifold to have a Lagrangian immersion into $\mathbf{C}^n$.

I.4.a   Lagrangian submanifolds described by functions

We consider firstly graphs.

Proposition I.4.1. *The graph of a map $F : \mathbf{R}^n \to (i)\mathbf{R}^n$ is a Lagrangian submanifold if and only if F is the gradient of a function $f : \mathbf{R}^n \to \mathbf{R}$.*

Proof. The tangent space to the graph at the point $(x, F(x))$ is the graph of $(dF)_x$, the differential of F at the point x. This graph is a Lagrangian subspace if and only if $(dF)_x$ is a symmetric endomorphism (see the proof of Lemma I.3.2). The matrix $\partial F_i/\partial x_j$ is symmetric for all x if and only if the differential form $\sum F_i dx_i$ over $\mathbf{R}^n$ is closed or, equivalently, exact:

$$F_i = \frac{\partial f}{\partial x_i}, \text{ namely } F = \nabla f.$$

$\square$

See, more generally, Proposition II.2.1.

The Lagrangian submanifolds obtained as graphs have a very specific property: the projection of the Lagrangian submanifold on $\mathbf{R}^n$ is a diffeomorphism. We would like to consider more general Lagrangian immersions, for instance immersions of compact manifolds. Here is a way to construct Lagrangian immersions using the reduction process of §I.2.b. We start from a Lagrangian submanifold[2] $L \subset \mathbf{C}^{n+k}$. We want to construct a Lagrangian immersion into $\mathbf{C}^n$. To write $\mathbf{C}^n$ as F/F°, we choose the co-isotropic subspace $F = \mathbf{C}^n \oplus \mathbf{R}^k$, the orthogonal of which is $F^\circ = 0 \oplus \mathbf{R}^k$. We suppose that the submanifold L is "transversal to F" in the sense that, for all x,

$$T_x L + F = \mathbf{C}^{n+k}.$$

The Lagrangian subspace $T_x L$ thus satisfies the assumption of the reduction lemma (Lemma I.2.3). Hence the composition

$$T_x L \cap F \subset F \longrightarrow F/F^\circ = \mathbf{C}^n$$

is the injection of a Lagrangian subspace.

Consider now the intersection V of the submanifold L with F. With the transversality assumption we have made on L, V is an n-dimensional submanifold of F (a consequence of the inverse function theorem) whose tangent space $T_x V$ is the intersection of $T_x L$ with F. Thus, the reduction lemma asserts, at the level of each tangent space, that, for all x in $V = L \cap F$, we have the injection of a Lagrangian subspace

$$T_x V \longrightarrow \mathbf{C}^n.$$

In other words, the composition

$$V = L \cap F \subset F \longrightarrow F/F^\circ = \mathbf{C}^n$$

is a Lagrangian immersion.

Remark I.4.2. Even if one starts from a Lagrangian submanifold, what we get in general is only an immersion.

Generating functions

We generalize the "graph" construction, using the reduction process as explained. Let us start with a nice and useful example.

Example I.4.3 *(The Whitney immersion)*. Consider the unit sphere in $\mathbf{R}^{n+1}$

$$S^n = \left\{ (x, a) \in \mathbf{R}^n \times \mathbf{R} \mid \|x\|^2 + a^2 = 1 \right\}$$

and the map

$$f : S^n \longrightarrow \mathbf{C}^n$$
$$(x, a) \longmapsto (1 + 2ia)x.$$

---

[2]Or a Lagrangian immersion.

The tangent space to the sphere is

$$T_{(x,a)}S^n = \{(\xi, \alpha) \in \mathbf{R}^n \times \mathbf{R} \mid x \cdot \xi + a\alpha = 0\}$$

and the tangent mapping to f is

$$T_{(x,a)}f : T_{(x,a)}S^n \longrightarrow \mathbf{C}^n$$
$$(\xi, \alpha) \longmapsto \xi + 2i(a\xi + \alpha x).$$

The map $T_{(x,a)}f$ is injective for all $(x,a) \in S^n$: if $T_{(x,a)}f(\xi, \alpha) = 0$, then $\xi = 0$ and $\alpha x = 0$; if $x = 0$, we have $a = \pm 1$ and the equality $x \cdot \xi + a\alpha = 0$ gives $\alpha = 0$. Thus we have $\xi = 0$ and $\alpha = 0$, so that f is an immersion. Moreover, we have

$$\omega(\xi + 2i(a\xi + \alpha x), \xi' + 2i(a\xi' + \alpha' x)) = 2(\xi \cdot (a\xi' + \alpha' x) - \xi' \cdot (a\xi + \alpha x))$$
$$= 2(\alpha' \xi \cdot x - \alpha \xi' \cdot x) = 0$$

so that the image of $T_{(x,a)}f$ is an isotropic subspace of dimension n, a Lagrangian subspace. In conclusion, the map f is a Lagrangian immersion. It has a unique double point (North and South poles of the sphere are mapped to 0). In dimension 1, this is a "figure eight". Below (in §I.4.b) we will draw pictures in dimensions 1 and 2.

Obviously, the Whitney sphere is not the graph of a map from $\mathbf{R}^n$ to $\mathbf{R}^n$. Let us show that it can nevertheless be described from the graph of a map defined on a larger space. We start from a function

$$f : \mathbf{R}^n \times \mathbf{R}^k \longrightarrow \mathbf{R}.$$

As we have seen it above, the graph of ∇f is a Lagrangian subspace of $\mathbf{C}^{n+k}$. We reduce $\mathbf{C}^{n+k}$ as in §I.2.b using the co-isotropic subspace $F = \mathbf{C}^n \oplus \mathbf{R}^k$. Here we intersect the graph of ∇f with F, namely we consider

$$V = \left\{ (x,a) \in \mathbf{R}^n \times \mathbf{R}^k \mid \frac{\partial f}{\partial a_1} = \cdots = \frac{\partial f}{\partial a_k} = 0 \right\} \subset \mathbf{R}^n \times \mathbf{R}^k.$$

The transversality assumption above is equivalent to the assumption that V is a submanifold of $\mathbf{R}^n \times \mathbf{R}^k$, in other words that the map

$$\mathbf{R}^n \times \mathbf{R}^k \longrightarrow \mathbf{R}^k$$
$$(x,a) \longmapsto \left(\frac{\partial f}{\partial a_1}, \ldots, \frac{\partial f}{\partial a_k} \right)$$

is a submersion along V. In terms of partial derivatives, this is to say that the matrix

$$\left(\left(\frac{\partial^2 f}{\partial a_i \partial a_j} \right)_{\substack{1 \le i \le k \\ 1 \le j \le k}} \left(\frac{\partial^2 f}{\partial x_i \partial a_j} \right)_{\substack{1 \le i \le n \\ 1 \le j \le k}} \right)$$

has maximal rank k. In terms of tangent subspaces, this is to say that the Lagrangian subspaces that are tangent to the graph of ∇f are transversal to the co-isotropic subspace F. The reduction lemma I.2.3 says that the map

$$V \longrightarrow \mathbf{R}^n \times \mathbf{R}^n = \mathbf{C}^n$$
$$(x, a) \longmapsto \left(x, \frac{\partial f}{\partial x_1}, \dots, \frac{\partial f}{\partial x_n} \right)$$

is a Lagrangian immersion.

Example I.4.4 *(The Whitney immersion, again).* With $k = 1$ and $f(x,a) = a \, \|x\|^2 + \dfrac{a^3}{3} - a$, we get

$$\frac{\partial f}{\partial a} = \|x\|^2 + a^2 - 1 = 0,$$

an equation which describes the sphere $S^n \subset \mathbf{R}^n \times \mathbf{R}$, and $\partial f / \partial x = 2ax$ gives the Whitney map.

Example I.4.5 *(Unfolding).* Unfoldings are deeply related with Lagrangian submanifolds (see [2]). I will not explain here the general theory but rather show an example. Let $P \in \mathbf{R}[X]$ be a degree-$(n + 1)$ polynomial

$$P(X) = X^{n+1} + x_1 X^{n-1} + \cdots + x_{n-1} X$$

where $x_1, \dots, x_{n-1} \in \mathbf{R}$. These coefficients are going to vary, this is the reason why they are named as variables. Call P_x the polynomial corresponding to $x = (x_1, \dots, x_{n-1}) \in \mathbf{R}^{n-1}$ and consider the map

$$f : \mathbf{R}^{n-1} \times \mathbf{R} \longrightarrow \mathbf{R}$$
$$(x_1, \dots, x_{n-1}, a) \longmapsto P_x(a)$$

to which we apply the previous techniques. The manifold V is

$$V = \left\{ (x, a) \in \mathbf{R}^{n-1} \times \mathbf{R} \mid \frac{\partial f}{\partial a}(x_1, \dots, x_{n-1}, a) = 0 \right\}$$
$$= \left\{ (x, a) \in \mathbf{R}^{n-1} \times \mathbf{R} \mid P'_x(a) = 0 \right\},$$

this is the set of critical points of P_x (zeroes of its derivative P'_x) when x varies. The condition that V actually be a submanifold is that the matrix of partial derivatives

$$\left(\left(\frac{\partial^2 f}{\partial a^2} \right), \left(\frac{\partial^2 f}{\partial x_i \partial a} \right)_{1 \leq i \leq n-1} \right)$$

has rank 1. But

$$\frac{\partial f}{\partial a} = P'_x(a) = (n+1)a^n + (n-1)x_1 a^{n-2} + \cdots + x_{n-1}$$

so that $\partial^2 f/\partial x_{n-1}\partial a$ is identically 1. Thus V is indeed a submanifold. The Lagrangian immersion is

$$V \longrightarrow T^\star \mathbf{R}^{n-1}$$

$$(x,a) \longmapsto \left(x, \frac{\partial P_x}{\partial x_1}(a), \ldots, \frac{\partial P_x}{\partial x_{n-1}}(a)\right).$$

For instance, starting from the family

$$P_x(X) = X^4 + x_1 X^2 + x_2 X,$$

we get

$$V = \left\{(x_1, x_2, a) \in \mathbf{R}^3 \mid 4a^3 + 2x_1 a + x_2 = 0\right\}$$

and the Lagrangian immersion from V into $\mathbf{R}^2 \times \mathbf{R}^2$ is the map

$$(x_1, x_2, a) \longrightarrow (x_1, x_2, a^2, a).$$

Figure I.1 shows V with its projection on the plane $\mathbf{R}^2$ of coefficients (x_1, x_2). The cusp curve is the discriminant of the family of degree-3 polynomials, the set of points x such that P'_x has a multiple root. It is obtained here as the set of critical values of the projection $V \to \mathbf{R}^2$. Over such a point x in the space of coefficients are the (one or three) roots of the polynomial P'_x.

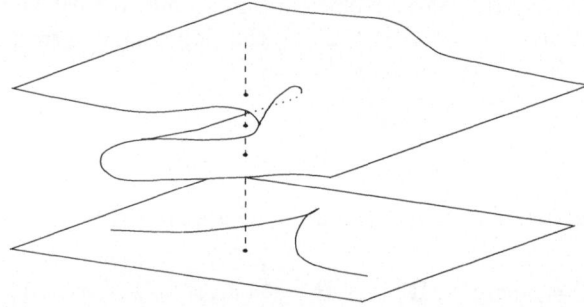

Figure I.1: The discriminant of degree-3 polynomials

I.4.b   Wave fronts

Exact Lagrangian immersions

If $f : V \to \mathbf{C}^n$ is a Lagrangian immersion, the 2-form $f^\star \omega$ is zero, so that $d(f^\star \lambda) = 0$ and $f^\star \lambda$ is a closed 1-form on V. If, for some reason, for instance because $H^1_{DR}(V) = 0$, this form is exact, there exists a function

$$F : V \longrightarrow \mathbf{R}$$

such that $f^\star \lambda = dF$. The immersion f is qualified as *exact Lagrangian immersion*. The mapping

$$F \times f : V \longrightarrow \mathbf{C}^n \times \mathbf{R}$$

has the property[3]

$$(f \times F)^\star \left(dz - \sum_{j=1}^n y_j dx_j \right) = 0.$$

Wave fronts

Instead of looking at the Lagrangian immersion f, consider the projection

$$V \xrightarrow{\ f \times F\ } \mathbf{C}^n \times \mathbf{R} \longrightarrow \mathbf{R}^n \times \mathbf{R}$$
$$(X + iY, z) \longmapsto (X, z).$$

We will assume here that, at a general point of the Lagrangian, the tangent space is transversal to the subspace of coordinates Y. The image of the Lagrangian immersion is then a hypersurface of $\mathbf{R}^n \times \mathbf{R}$. This hypersurface is the *wave front*. Of course, it will in general be singular. Precisely, at a point of V where V is not a graph over $\mathbf{R}^n$, the projection $X + iY \mapsto X$ is singular. However, as $(f \times F)^\star \left(dz - \sum_{j=1}^n y_j dx_j \right) = 0$, at every point of the wave front, there is a tangent hyperplane, the hyperplane

$$z = \sum_{j=1}^n Y_j x_j \text{ in the space } \mathbf{R}^n \times \mathbf{R} \text{ of } (x, z) \text{ coordinates}$$

at the point image of (X, Y, z). Notice that, as the coefficient of z in this equation is non zero, the hyperplane is always transversal to the z-axis. Conversely, if a singular hypersurface of $\mathbf{R}^n \times \mathbf{R}$ has at every point a tangent hyperplane that is transversal to the z-axis, this hyperplane has a unique equation of the form $z = \sum Y_j x_j$ and it is possible to reconstruct a (maybe singular) Lagrangian submanifold from the "slopes" Y_j.

We begin with an example of dimension 1, that of the Whitney immersion again. Notice that this is indeed an exact Lagrangian immersion: the restriction of the Liouville form ydx to the curve is exact because $\int ydx = 0$ (the "algebraic" area surrounded by the curve is zero). A primitive of ydx is easily found. The curve is parametrized by $t \mapsto (\cos t, \sin 2t)$ and

$$ydx = -2 \sin^2 t \cos t = -\frac{2}{3} d \left(\sin^3 t \right).$$

---

[3]The manifold $\mathbf{C}^n \times \mathbf{R}$ is a "contact manifold" and $F \times f$ is a "Legendrian immersion" lifting the Lagrangian immersion F.

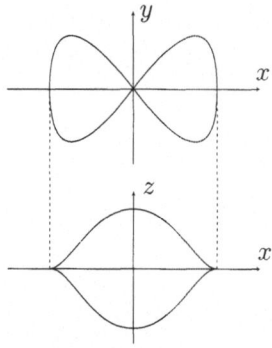

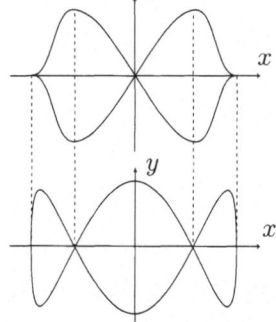

Figure I.2: Eye   Figure I.3: Crossbow

A map to the (x, z) space is thus

$$t \mapsto (x, z) = \left(\cos t, -\frac{2}{3} \sin^3 t \right).$$

This is depicted on Figure I.2, in an old-fashion "descriptive geometry" mood. It can be seen that the singular points of the (x, z) curve correspond to the tangents to the (x, y) curve that are vertical, and that the double point of the latter corresponds to the two tangents to the wavefront (the "eye") at points with the same x coordinate that are parallel.

Figure I.3 represents an example in which we start from the wave front (a "crossbow") to reconstruct the Lagrangian. From the wave front, it is seen that the Lagrangian curve has two double points and two "vertical" tangents.

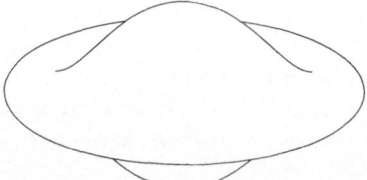

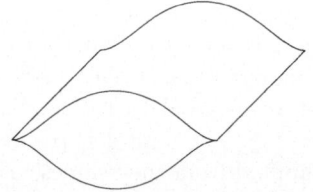

Figure I.4: Flying saucer   Figure I.5: Cylinder

One could wonder what it is useful for to replace an immersed curve by a singular one. Notice that, in higher dimensions, the wave front is a hypersurface in $\mathbf{R}^n \times \mathbf{R}$ and it replaces a submanifold of the same dimension n in $\mathbf{R}^n \times \mathbf{R}^n$. Even for $n = 2$, this is very useful as this allows to represent exact Lagrangian surfaces of $\mathbf{R}^4$ by (singular) surfaces in a dimension-3 space. Here are some beautiful examples. Rotate the eye (Figure I.2) about the z-axis to get the flying saucer depicted on Figure I.4. The corresponding Lagrangian surface in $\mathbf{R}^2 \times \mathbf{R}^2$ is a Lagrangian immersion of the dimension-2 sphere in $\mathbf{C}^2$ with a double point. In Exercise I.14, one checks that this is, indeed, the Whitney immersion... eventually drawn in dimension 2!

Figure I.5 represents a cylinder constructed on the eye, namely a Lagrangian immersion of a cylinder, product of a figure eight with an interval, with two whole lines of singular points.

Singularities

Wave fronts are, as we have said it, singular hypersurfaces. We have seen, in dimension 1, cusps, in dimension 2, lines of cusps, but this can be more complicated, as Exercise I.15 shows it.

Wave fronts of non exact Lagrangian immersions

Wave fronts are so nice that it is a pity not to have them for all Lagrangian immersions. In dimension 1, the problem is to represent by wave fronts curves that do not surround a zero area. Consider for instance the standard (round) circle in $\mathbf{C}$. As $\int y \, dx \neq 0$, it seems that nothing can be done. Look, however, at the parametrization

$$t \longmapsto (\cos t, \sin t).$$

It gives

$$y \, dx = -\sin^2 t \, dt = d\left(\frac{\sin 2t}{4} - \frac{t}{2} + C\right).$$

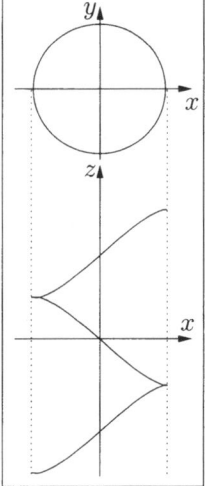

Figure I.6: Wave front of the circle

Nothing forbids us to represent the Lagrangian (non exact) immersion of the circle by a piece of the (non closed) wave front[4] parametrized by

$$t \longmapsto \left(\cos t, \frac{\sin 2t}{4} - \frac{t}{2} + C \right)$$

and depicted on Figure I.6[5].

Figure I.7

If we rotate the (unbounded) wave front of Figure I.6 around a line parallel to the z-axis that does not intersect the wave front, we get the wave front of a Lagrangian torus, the one depicted on Figure I.7. One can then use the cylinder represented on Figure I.5 to perform connected sums of wave fronts. This way, Figure I.8 represents (the wave front of) a genus-2 Lagrangian surface. In the same way, one constructs Lagrangian immersions of all orientable surfaces in $\mathbf{C}^2$. These figures are copied from Givental's paper [13], that contains many other examples.

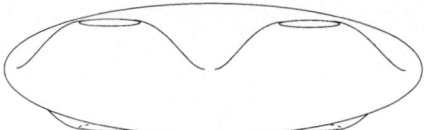

Figure I.8: A genus-2 surface

Remark I.4.6. Except for the torus, all the surfaces depicted here have double points, that show up in the wave fronts as points having the same projection on the horizontal plane and parallel tangent planes. It is rather easy to prove that the torus is the only orientable surface that can be *embedded* as a Lagrangian submanifold in $\mathbf{C}^2$. As for non orientable surfaces, they can be embedded as Lagrangian surfaces when (and only when) their Euler characteristic is divisible by 4 (with the exception of the Klein bottle). See the pictures in [13]. As for the Klein bottle, it has long been unknown whether it had or had not a Lagrangian embedding. Mohnke [27] has recently proved that it has not.

---

[4]This is a place where one can really appreciate the difference between closed and exact 1-forms.

[5]Notice that wave fronts are defined only up to a "vertical" translation, the actual constant C used in Figure I.6 is $(\pi + 1)/4$.

Exact Lagrangian embeddings

Notice that, in all the examples of exact Lagrangian immersions we have given, there are double points. This is obviously necessary in dimension 2 ($n = 1$), due to Jordan theorem: an embedded curve cannot surround a zero area. This is also true in higher dimensions, due to a (hard) theorem of Gromov [16]: there is no exact Lagrangian submanifold in $\mathbf{C}^n$.

I.4.c   Other examples

Here are a few other examples.

Grassmannians

Consider the map

$$\mathrm{U}(n) \longrightarrow \mathrm{Sym}(n; \mathbf{C})$$
$$A \longmapsto {}^t\!AA$$

from the group $\mathrm{U}(n)$ to the complex vector space of symmetric matrices.

Proposition I.4.7. *The map* $A \mapsto {}^t\!AA$ *defines a Lagrangian immersion*

$$\Phi : \Lambda_n \longrightarrow \mathrm{Sym}(n; \mathbf{C}).$$

Proof. As ${}^t\!AA = \mathrm{Id}$ when $A \in \mathrm{O}(n)$, the map Φ is well defined. Call $[A]$ the class of a unitary matrix A in Λ_n. We have seen in §I.3.d that the tangent space to Λ_n at the point $[A]$ can be identified with

$$T_{[A]}\Lambda_n = \{AH \mid H \in i\,\mathrm{Sym}(n; \mathbf{R})\}.$$

It is mapped into $\mathrm{Sym}(n; \mathbf{C})$ par $T_{[A]}\Phi$ as follows

$$AH \longmapsto {}^t\!A\left(AH^t\bar{A} + \bar{A}^t H^t A\right) A.$$

The matrix $AH^t\bar{A} + \bar{A}^t H^t A$ has the form $K - \bar{K}$ for $K = AH^t\bar{A} = AHA^{-1}$ in $\mathfrak{u}(n)$ and this describes all the matrices in the vector space $i\,\mathrm{Sym}(n; \mathbf{R})$ when H varies in $i\,\mathrm{Sym}(n; \mathbf{R})$. The image of the tangent mapping $T_{[A]}\Phi$ is, thus, the subspace $\rho(A) \cdot i\,\mathrm{Sym}(n; \mathbf{R})$ where

$$\rho : \mathrm{U}(n) \longrightarrow \mathrm{U}\left(\frac{n(n+1)}{2}\right)$$
$$A \longmapsto (B \mapsto {}^t\!ABA)$$

is the representation of $\mathrm{U}(n)$ operating on complex symmetric matrices. This image is, indeed, a Lagrangian subspace, being the image of the real part of the complex vector space $i\,\mathrm{Sym}(n; \mathbf{C})$ by a unitary matrix. $\qquad\square$

Tori, integrable systems

Integrable systems (mechanical systems with many conserved quantities) yield many Lagrangian tori. We use here a few standard symplectic notions: Hamiltonian vector fields, Poisson bracket, commuting functions. See if necessary Appendix I.6.c. Recall for instance that an integrable system on $\mathbf{C}^n = \mathbf{R}^{2n}$ is a map $f : \mathbf{R}^{2n} \to \mathbf{R}^n$ whose components $f_1, \ldots, f_n$ are functionally independent commuting functions.

   This defines a local $\mathbf{R}^n$-action on $\mathbf{R}^{2n}$, which is locally free at the regular points of the system (the points at which the derivatives of the functions f_i are actually independent). Call $X_1, \ldots, X_n$ the Hamiltonian vector fields associated with the functions f_i. These vector fields commute:

$$[X_i, X_j] = X_{\{f_i, f_j\}} = 0.$$

The $\mathbf{R}^n$-action is given by integration:

$$t \cdot x = \varphi_n^{t_n} \circ \varphi_{n-1}^{t_{n-1}} \circ \cdots \circ \varphi_1^{t_1}(x)$$

where φ_i denotes the flow of X_i and $t = (t_1, \ldots, t_n) \in \mathbf{R}^n$ is close to 0 (in order that $\varphi_i^{t_i}$ be defined). This local action is indeed locally free on the open set of regular points because the vector fields X_i give independent tangent vectors at these points.

   Assume moreover that the vector fields X_i are complete, namely that the flows $\varphi_i^{t_i}$ are defined for all values of t. We then have a locally free action of $\mathbf{R}^n$ on the whole set of regular points. The vector fields X_i being tangent to the common level sets of the f_i's, this action preserve the level sets. The connected components of the regular level sets of f are thus homogeneous spaces, quotients of $\mathbf{R}^n$ by discrete subgroups. The discrete subgroups of $\mathbf{R}^n$ are the lattices $\mathbf{Z}^k$ in the linear subspaces of dimension k. The connected components of the regular level sets are thus diffeomorphic to $\mathbf{R}^{n-k} \times \mathbf{T}^k$ for some k such that $0 \leq k \leq n$. In particular, the compact connected components are tori $\mathbf{T}^n$ and these tori are Lagrangian[6], they are called the *Liouville tori*. The next proposition is the easiest part of the Arnold-Liouville theorem (see for instance [2, 6]).

Proposition I.4.8. *Compact connected components of the regular common level sets of an integrable system are Lagrangian tori.*   □

   There are many examples of integrable systems and thus of Lagrangian tori, coming from mechanical systems (spinning top, pendulum...)[7]. The most classical example is that of the standard action of the torus

$$T^n = \{(t_1, \ldots, t_n) \in \mathbf{C}^n \mid |t_j| = 1, \quad i = 1, \ldots, n\}$$

---

[6]Notice that on a compact connected component, the flows *are* complete.
[7]See for instance [6].

on $\mathbf{C}^n$ by

$$(t_1, \ldots, t_n) \cdot (z_1, \ldots, z_n) = (t_1 z_1, \ldots, t_n z_n),$$

the orbits of which are the common level sets of the functions

$$g_1 = \frac{1}{2}|z_1|^2, \ldots, g_n = \frac{1}{2}|z_n|^2,$$

tori $S^1 \times \cdots \times S^1$ indeed, for the regular values of the g_i's (namely every g_i non zero). We will come back to these examples in §I.5.e.

Normal bundles

Let now $f : V \to \mathbf{R}^n$ be any immersion of a k-dimensional manifold into $\mathbf{R}^n$. Consider the total space of its normal bundle

$$Nf = \left\{ (x, v) \in V \times \mathbf{R}^n \mid x \in V, \quad v \in (T_x f(T_x V))^\perp \right\}.$$

It is naturally mapped into $\mathbf{R}^n \times \mathbf{R}^n$ by

$$\begin{aligned} \widetilde{f} : \quad Nf &\longrightarrow \mathbf{R}^n \times \mathbf{R}^n \\ (x, v) &\longmapsto (f(x), v). \end{aligned}$$

The manifold Nf has dimension $k + n - k = n$, and $\widetilde{f}$ is clearly an immersion. Moreover, it is Lagrangian. More precisely, we have:

Lemma I.4.9. *If λ is the Liouville form on $\mathbf{R}^n \times \mathbf{R}^n$, one has $\widetilde{f}^\star \lambda = 0$.*

Proof. Consider a vector $X \in T_{(x,v)} Nf$. Use the commutative diagram

$$\begin{array}{ccc} Nf & \xrightarrow{\;\widetilde{f}\;} & \mathbf{R}^n \times \mathbf{R}^n \\ {\scriptstyle \pi}\Big\downarrow & & \Big\downarrow{\scriptstyle \pi} \\ V & \xrightarrow{\;f\;} & \mathbf{R}^n \end{array}$$

to compute

$$\begin{aligned} \left(\widetilde{f}^\star \lambda \right)_{(x,v)} (X) &= \lambda_{(f(x),v)} \left(T_{(x,v)} \widetilde{f}(X) \right) \\ &= v \cdot \left(T_{(f(x),v)} \pi \circ T_{(x,v)} \widetilde{f}(X) \right) \\ &= v \cdot \left(T_x f \circ T_{(x,v)} \pi(X) \right) \\ &= 0 \text{ since } v \text{ is orthogonal to } T_x f(T_x V). \end{aligned}$$

$\square$

This method allows to construct many (non compact) examples and can be generalized by replacing $\mathbf{R}^n \times \mathbf{R}^n = T^\star \mathbf{R}^n$ by the cotangent bundle $T^\star M$ of a manifold and $V \to \mathbf{R}^n$ by an immersion into M. See §II.2.a.

I.4.d   The Gauss map

Let $f : V \to \mathbf{C}^n$ be a Lagrangian immersion. Its tangent space at any point is a Lagrangian subspace of $\mathbf{C}^n$. One can globalize the data consisting of all these tangent spaces to define the "Gauss map"

$$V \xrightarrow{\ \gamma(f)\ } \Lambda_n$$
$$x \longmapsto T_x f(T_x V).$$

By definition of the tautological bundle (§ I.3.c), one has

$$\gamma(f)^* E_n = TV.$$

In particular, the tangent bundle to V must have the same properties as E_n.

Proposition I.4.10. *For a manifold to have a Lagrangian immersion into $\mathbf{C}^n$, it is necessary that the complexification of its tangent bundle be trivializable.* $\square$

The converse is true, but less easy to prove. This is an application of Gromov's h-principle [17], see also [23].

Examples I.4.11.   (1) Spheres. We have seen examples of Lagrangian immersions of spheres in $\mathbf{C}^n$ (in § I.4.a). One deduces that $TS^n \otimes_{\mathbf{R}} \mathbf{C}$ is a trivial complex bundle. Notice however that it is not true that the tangent bundle TS^n itself is trivial (except for $n = 0$, 1, 3 and 7).

(2) Surfaces. All orientable surfaces and half the non orientable surfaces have Lagrangian immersions in $\mathbf{C}^2$ (as we have seen it in § I.4.b). This is not the case, neither for the real projective plane nor for the connected sums of an odd number of copies of this plane.

(3) Normal bundles. This is a case where the tangent bundle itself is trivial (before complexification):

$$T_{(x,v)}(Nf) = \{(\xi, U) \mid \xi \in T_x V, U \perp T_x f(T_x V)\}$$
$$= T_x V \oplus N_x f$$

and this is canonically isomorphic to the ambient space $\mathbf{R}^n$.

(4) Grassmannians. The Gauss map φ of the Lagrangian immersion Φ

$$\varphi : \Lambda_n \longrightarrow \Lambda_{\frac{n(n+1)}{2}}$$

satisfies of course

$$\varphi^* E_{\frac{n(n+1)}{2}} = \operatorname{End} \operatorname{Sym}(E_n).$$

The Maslov class

Every Lagrangian immersion has a Maslov class: use the Gauss map

$$\varphi(f) : V \longrightarrow \Lambda_n$$

to pull back $\mu \in H^1(\Lambda_n; \mathbf{Z})$ to a class

$$\mu(f) \in H^1(V; \mathbf{Z}).$$

One can also, with the notation of Remark I.3.3, define $\mu(f)$ as the cohomology class dual to $\gamma(f)^{-1}(\Sigma_n)$, see [24] for example.

I.5 Special Lagrangian submanifolds in $\mathbf{C}^n$

Lagrangian submanifolds are submanifolds of $\mathbf{C}^n$ whose tangent space at each point is a Lagrangian subspace. They have a Gauss map into the Grassmannian Λ_n, namely into $\mathrm{U}(n)/\mathrm{O}(n)$. We look now at the submanifolds whose Gauss map takes values in $S\Lambda_n = \mathrm{SU}(n)/\mathrm{SO}(n)$. These are the special Lagrangian submanifolds, invented by Harvey and Lawson [18].

I.5.a Special Lagrangian subspaces

An oriented subspace P of $\mathbf{C}^n$ is said to be *special Lagrangian* if it has a positive orthonormal basis that is a *special unitary* basis of $\mathbf{C}^n$.

For instance, if $n = 1$, as $\mathbf{C}$ has a unique special unitary basis (the group $\mathrm{SU}(1)$ is the trivial group), there is only one special Lagrangian subspace in $\mathbf{C}$, the line $\mathbf{R} \subset \mathbf{C}$... this will not be a very interesting notion in dimension 1. Fortunately, for $n \geq 2$, this is more exciting. Identify the space $\mathbf{C}^2$ with the skew-field $\mathbf{H}$ of quaternions:

$$\begin{aligned} Z = (z_1, z_2) &= X + iY \\ &= (x_1 + iy_1, x_2 + iy_2) \\ &= (x_1 + iy_1) + j(x_2 + iy_2) \\ &= (x_1 + jx_2) + i(y_1 - jy_2). \end{aligned}$$

The 2×2 matrices that are in $\mathrm{SU}(2)$ are the matrices of the form

$$\begin{pmatrix} z_1 & -\bar{z}_2 \\ z_2 & \bar{z}_1 \end{pmatrix} \quad \text{with } |z_1|^2 + |z_2|^2 = 1.$$

Thus the special Lagrangian planes are those who have an orthonormal basis (Z, Z') with Z and Z' of the form

$$\begin{cases} Z = (x_1 + iy_1) + j(x_2 + iy_2) \\ Z' = (-x_2 + iy_2) + j(x_1 - iy_1). \end{cases}$$

Notice that
$$Z' = [(x_1 + iy_1) + j(x_2 + iy_2)]\, j = Zj.$$

Thus a basis (Z, Z') of $\mathbf{C}^2$ is special unitary if and only if $Z' = Zj$. Now use multiplication by j to give $\mathbf{H}$ the structure of a complex vector space. One has:

Proposition I.5.1. *The special Lagrangian subspaces of $\mathbf{C}^2$ are the complex lines with respect to the complex structure defined by the multiplication by j. The Grassmannian $S\Lambda_2$ is a complex projective line.* $\qquad\qquad\qquad\square$

Remark I.5.2. Notice also that $S\Lambda_2 = \mathrm{SU}(2)/\mathrm{SO}(2) = S^3/S^1$. This is indeed a dimension-2 sphere.

To distinguish the special Lagrangian subspaces among all the Lagrangian subspaces or the special unitary matrices among all the unitary matrices, one uses the (complex) determinant. To globalize the notion of special Lagrangian subspace and define special Lagrangian submanifolds, it will be practical (and natural) to describe the linear objects by differential forms. The form corresponding to the complex determinant is
$$\Omega = dz_1 \wedge \cdots \wedge dz_n.$$

Expressing the definition of the determinant, namely
$$(Ae_1) \wedge \cdots \wedge (Ae_n) = (\det A)e_1 \wedge \cdots \wedge e_n,$$

we see that, for $A \in \mathrm{GL}(n; \mathbf{C})$, we have indeed
$$A^\star \Omega = (\det A)\Omega.$$

Hence
$$\det A = 1 \iff A^\star \Omega = \Omega.$$

In order to work with real subspaces, we need an additional notation: call α and β the two degree n real forms:
$$\alpha = \operatorname{Re}\Omega, \qquad \beta = \operatorname{Im}\Omega.$$

For instance, in dimension 1, $\Omega = dz$, $\alpha = dx$ and $\beta = dy$. In dimension 2,
$$\begin{aligned}
\Omega = dz_1 \wedge dz_2 &= (dx_1 + idy_1) \wedge (dx_2 + idy_2) \\
&= dx_1 \wedge dx_2 - dy_1 \wedge dy_2 + i(dy_1 \wedge dx_2 + dx_1 \wedge dy_2),
\end{aligned}$$

that is
$$\begin{cases} \alpha = dx_1 \wedge dx_2 - dy_1 \wedge dy_2 \\ \beta = dy_1 \wedge dx_2 + dx_1 \wedge dy_2. \end{cases}$$

Proposition I.5.3. *Let P be an oriented (real) vector subspace of dimension n in $\mathbf{C}^n$. The number $\Omega(x_1 \wedge \cdots \wedge x_n)$ depends only on P and not on the positive orthonormal basis $(x_1, \ldots, x_n)$ of P used to express it.*

Proof. Consider the $2n$ vectors $(x_1, \ldots, x_n, ix_1, \ldots, ix_n)$ and the linear mapping $A : \mathbf{C}^n \to \mathbf{C}^n$ defined by the images of the vectors of the canonical basis:

$$A(e_j) = x_j, \quad A(ie_j) = ix_j$$

(so that A is complex linear). Then

$$\Omega(x_1 \wedge \cdots \wedge x_n) = \det_{\mathbf{C}} A.$$

If $(gx_1, \ldots, gx_n)$ is a positive orthonormal basis of P (that is, if $g \in SO(n)$), one gets

$$\Omega(gx_1 \wedge \cdots \wedge gx_n) = \det_{\mathbf{C}}(gA) = \det_{\mathbf{C}} g \det_{\mathbf{C}} A = \det_{\mathbf{R}} g \det_{\mathbf{C}} A$$
$$= \det_{\mathbf{C}} A = \Omega(x_1 \wedge \cdots \wedge x_n)$$

(since $g \in SO(n) \subset GL(n; \mathbf{R}) \subset GL(n; \mathbf{C})$). $\qquad\square$

We will thus denote $\Omega(P)$ the number $\Omega(x_1 \wedge \cdots \wedge x_n)$. Similarly, denote $\alpha(P)$ and $\beta(P)$ its real and imaginary parts.

Remark I.5.4. Notice that $\Omega(P)$ is non zero if and only if the $2n$ vectors

$$(x_1, \ldots, x_n, ix_1, \ldots, ix_n)$$

form a basis of $\mathbf{C}^n$ over $\mathbf{R}$, that is, if and only if $P \cap iP = \{0\}$ or P does not contain any complex line. These subspaces are said to be *totally real*. This is in particular the case for Lagrangian subspaces.

Proposition I.5.5. *A real subspace P of $\mathbf{C}^n$ has an orientation for which it is a special Lagrangian subspace if and only if P is Lagrangian and $\beta(P) = 0$.*

Proof. Let P be a Lagrangian subspace. Choose $(x_1, \ldots, x_n)$, an orthonormal basis which is the image of the canonical basis of $\mathbf{C}^n$ by a unitary matrix A. Thus

$$\Omega(P) = \det_{\mathbf{C}} A \in S^1.$$

For P to have a positive basis that is special unitary, it is necessary and sufficient that $\det_{\mathbf{C}} A$ be equal to ± 1, that is, that

$$\operatorname{Im} \Omega(P) = 0.$$

$\qquad\square$

Here is a last elementary remark on linear subspaces:

Proposition I.5.6. *Let $Q \subset \mathbf{C}^n$ be an oriented isotropic linear subspace of dimension $n - 1$. There exists a unique special Lagrangian subspace that contains Q.*

Proof. Choose a positive orthonormal basis $(x_1, \ldots, x_{n-1})$ of Q. In the complex line that is the orthogonal, with respect to the Hermitian form, of the complex subspace spanned by the x_i's, there is a unique vector x_n such that the basis $(x_1, \ldots, x_{n-1}, x_n)$ is a special unitary basis of $\mathbf{C}^n$. $\qquad\square$

I.5.b   Special Lagrangian submanifolds

A Lagrangian immersion

$$f : V \longrightarrow \mathbf{C}^n$$

of an oriented manifold into $\mathbf{C}^n$ is *special* if $T_x f(T_x V)$ is a special Lagrangian subspace for every x. The Gauss map then takes values in $S\Lambda_n \subset \tilde{\Lambda}_n$.

Examples I.5.7.   (1) In dimension 1, the tangent space must be the unique special Lagrangian $\mathbf{R} \subset \mathbf{C}$ for all x. If V is connected, f must thus be the immersion of an open subset of $\mathbf{R}$ by $t \mapsto t + ia$. We have already noticed that this dimension will not be very exciting.

(2) In dimension 2, $T_x f(T_x V)$ must be a j-complex line for all x, f is thus the immersion of a j-complex curve into $\mathbf{C}^2$. This gives quite a lot of examples.

Remark I.5.8. The Maslov class of a *special* Lagrangian immersion into $\mathbf{C}^n$ is zero. Of course, as the examples above show it, there are much more Lagrangian immersions with zero Maslov class than there are special Lagrangian immersions.

In terms of forms, to say that the immersion

$$f : V \longrightarrow \mathbf{C}^n$$

is special Lagrangian is to say that it satisfies

- firstly $f^\star \omega = 0$ (it is Lagrangian)

- secondly $f^\star \beta = 0$ (it is special).

Proposition I.5.9. *If f is a special Lagrangian immersion, $f^\star \Omega$ is a volume form on V.*

Proof. The complex form Ω has type $(n, 0)$ and defines an n-form $f^\star \Omega$ on V, which is real since its imaginary part vanishes on V. Let x be a point in V and let $(X_1, \ldots, X_n)$ be a basis of $T_x V$. One has

$$(f^\star \Omega)_x (X_1, \ldots, X_n) = \Omega_{f(x)}(T_x f(X_1), \ldots, T_x f(X_n)) \neq 0$$

because of Remark I.5.4 and since V is Lagrangian. Thus $f^\star \Omega$ never vanishes.   □

In dimensions 1 and 2, the special Lagrangian submanifolds are non compact (in dimension 2, Liouville's theorem forbids complex curves in $\mathbf{C}^2$ to be compact). This is actually always the case, a straightforward application of Proposition I.5.9:

Corollary I.5.10. *There is no special Lagrangian immersion from a compact manifold into $\mathbf{C}^n$.*

Proof. If $f : V \to \mathbf{C}^n$ is a special Lagrangian immersion, $f^\star \Omega$ is a volume form on V. But Ω is an exact complex form :

$$\Omega = dz_1 \wedge \cdots \wedge dz_n = d(z_1 dz_2 \wedge \cdots \wedge dz_n).$$

Decompose $z_1 dz_2 \wedge \cdots \wedge dz_n$ into its real and imaginary parts to get

$$\alpha = d \operatorname{Re}(z_1 dz_2 \wedge \cdots \wedge dz_n) = d\eta$$

and eventually

$$f^\star \Omega = f^\star \alpha = d(f^\star \eta).$$

The manifold V thus has an exact volume form, and this prevents it of being compact. $\qquad\square$

Let us give now examples of special Lagrangian submanifolds in $\mathbf{C}^n$, starting from the examples of Lagrangians constructed in section I.4.

I.5.c Graphs of forms

Let us begin with Proposition I.4.1. Let $f : \mathbf{R}^n \to \mathbf{R}$ be a function. We require the graph of ∇f, a Lagrangian submanifold, to be a special Lagrangian submanifold. The $n = 1$ case is not interesting. For $n = 2$, the Lagrangian immersion associated with the function f is

$$F : (x, y) \longmapsto \left(x, y, \frac{\partial f}{\partial x}, \frac{\partial f}{\partial y} \right)$$

and the form β is

$$\beta = dy_1 \wedge dx_2 + dx_1 \wedge dy_2.$$

Then

$$F^\star \beta = d\left(\frac{\partial f}{\partial x} \right) \wedge dy + dx \wedge d\left(\frac{\partial f}{\partial y} \right)$$

$$= \left(\frac{\partial^2 f}{\partial x^2} + \frac{\partial^2 f}{\partial y^2} \right) dx \wedge dy.$$

We thus have:

Proposition I.5.11. *Let U be an open subset of $\mathbf{R}^2$ and $f : U \to \mathbf{R}$ a function of class $\mathcal{C}^2$. The graph of ∇f is a special Lagrangian submanifold of $\mathbf{C}^2$ if and only if f is a harmonic function.* $\qquad\square$

Notice that the condition is linear. Starting from dimension 3, this is no more the case. The function f must satisfy a complicated non linear partial differential equation, expressed in Proposition I.5.12 below. Let us begin by a notation. Denote by Hess(f) the Hessian matrix of f, namely the matrix

$$\text{Hess}(f)_{i,j} = \frac{\partial^2 f}{\partial x_i \partial x_j}$$

and by $\sigma_k(\text{Hess}(f))$ the k-th elementary symmetric functions of its eigenvalues. More generally, for an $n \times n$ real matrix A, write

$$\det (A - X \, \text{Id}) = \sum_{k=0}^{n} (-1)^k \sigma_k(A) X^{n-k}.$$

For example, $\sigma_1(\text{Hess}(f))$ is the trace of the Hessian matrix, the Laplacian Δf of f.

Proposition I.5.12. *Let U be an open subset of $\mathbf{R}^n$ and $f : U \to \mathbf{R}$ a function of class $\mathcal{C}^2$. The graph of ∇f is a special Lagrangian submanifold of $\mathbf{C}^n$ if and only if f satisfies the partial differential equation*

$$\sum_{k \geq 0} (-1)^k \sigma_{2k+1}(\text{Hess}(f)) = 0.$$

Examples I.5.13. For $n = 1$, the differential equation is $f''(t) = 0$ or $f'(t)$ constant, and this is precisely the differential equation of the special Lagrangian submanifolds. For $n = 2$, again, only σ_1 appears in the (linear) relation, which expresses the fact that the function f must be harmonic. For $n = 3$, the relation is

$$\sigma_1(\text{Hess}(f)) = \sigma_3(\text{Hess}(f))$$

or

$$\Delta f = \det(\text{Hess}(f)).$$

Remark I.5.14. The only order-1 term (in df) in this partial differential equation is Δf, so that the "linear part" of this equation is $\Delta f = 0$. this should be compared with McLean's theorem (Theorem II.6.1 below).

Proof of the proposition. The tangent space to the graph of ∇f at the point $(x, \nabla f_x)$ is the image of the plane $\mathbf{R}^n$ under the linear map $\text{Id} + i(d^2 f)_x$. This is a special Lagrangian subspace if and only if

$$\text{Im} \left(\det_{\mathbf{C}}(\text{Id} + i(d^2 f)_x) \right) = 0.$$

We still must check that, for any real symmetric matrix A, one has

$$\text{Im} \left(\det_{\mathbf{C}}(\text{Id} + iA) \right) = \sum_{k \geq 0} (-1)^k \sigma_{2k+1}(A).$$

Since A is real symmetric, it is diagonalizable in an orthonormal basis. It is clear that the two sides of the relation to be proved are invariant under conjugation by matrices in $O(n)$. One may thus assume that the matrix A is the diagonal $(\lambda_1, \ldots, \lambda_n)$. The left hand side is then $\text{Im} \prod_j (1 + i\lambda_j)$ and it clearly coincides with the right hand side. $\qquad\square$

I.5.d Normal bundles of surfaces

Let $f : V \to \mathbf{R}^n$ be an immersion of a dimension-k manifold into $\mathbf{R}^n$. We know (see §I.4.c) that its normal bundle has a natural Lagrangian immersion into $\mathbf{R}^n \times \mathbf{R}^n$. Look now for the conditions under which this is a special Lagrangian immersion.

For the sake of simplicity, suppose here that $k = 2$ and $n = 3$ (case of surfaces in $\mathbf{R}^3$). There is a more general discussion in [18].

Fix a point x_0 in V, a unit normal vector field $n = n(x)$ on a neighbourhood of x_0. The restriction of the normal bundle

$$Nf = \{(x, v) \mid x \in V, v \in \mathbf{R}^3, v \perp T_x f(T_x v)\}$$

to this neighborhood is isomorphic with $V \times \mathbf{R}$ by $(x, \mu) \mapsto (x, \mu n(x))$. We map Nf to $\mathbf{C}^3$ by

$$(x, \mu) \mapsto \mu n(x) + i f(x)$$

(notice that, this time, the immersion f appears in the second copy of $\mathbf{R}^n$, that of purely imaginary vectors).

Let us now choose an orthonormal basis (e_1, e_2) of $T_{x_0} V$. Assume that this basis is orthogonal with respect to the second fundamental form, that is, to the symmetric bilinear form defined on $T_{x_0} V$ by

$$\mathrm{II}(X, Y) = -(T_{x_0} n(X), Y).$$

We have

$$T_{x_0} n(e_1) = -\lambda_1 e_1, \quad T_{x_0} n(e_2) = -\lambda_2 e_2$$

where λ_1 and λ_2 are the two "principal curvatures" of V at x_0.

Consider now the tangent space to Nf at (x_0, v) where $v = \mu n(x_0) \in N_{x_0} f = \mathbf{R} \cdot n(x_0)$. The tangent mapping to our immersion is

$$P_0 = T_{(x_0, \mu)}(Nf) = T_{x_0} V \oplus N_{x_0} f \longrightarrow \mathbf{R}^3 \times \mathbf{R}^3$$
$$(\xi, \eta) \longmapsto (\eta n(x_0) + \mu T_{x_0} n(\xi), T_{x_0} f(\xi)).$$

The images of the basis vectors are

$$e_1 \longmapsto (-\mu \lambda_1 e_1, e_1)$$
$$e_2 \longmapsto (-\mu \lambda_2 e_2, e_2)$$
$$n \longmapsto (n, 0).$$

Thus

$$\Omega(P_0) = (dz_1 \wedge dz_2 \wedge dz_3)\left(((i - \mu\lambda_1)e_1) \wedge ((i - \mu\lambda_2)e_2) \wedge n\right)$$
$$= (i - \mu\lambda_1)(i - \mu\lambda_2),$$

so that P_0 is a special Lagrangian if and only if $\mu(\lambda_1 + \lambda_2) = 0$. This is to say that the trace of $T_{x_0} n$ is zero. In other words, we have shown:

Proposition I.5.15. *The immersion of the normal bundle of*

$$f : V \longrightarrow \mathbf{R}^3$$

into **C**^{3} *is a special Lagrangian immersion if and only if f is a minimal immersion.*

$\square$

For more information on minimal surfaces, see, for example, the beautiful surveys in [29] and the references quoted there.

Remark I.5.16. It is true that we have already mentioned Riemannian metrics in these notes, but up to now, they have had only an auxiliary role. The result presented here is a genuine Riemannian one.

I.5.e From integrable systems

Being compact, Lagrangian tori obtained as "Liouville tori" cannot be special Lagrangian submanifolds in **C**^{n}. One can try to replace them by special Lagrangian submanifolds with the help of the remark included in Proposition I.5.6: the idea is to consider a (necessarily isotropic) subtorus in a Liouville torus T^n and to add a direction to construct another Lagrangian submanifold, which will be special.

Here is an example, coming from [18], of such a construction. Start from an orbit L of the standard action of T^n on **C**^{n} (see §I.4.c), namely a common level set of the functions

$$g_1(z) = \frac{1}{2} |z_1|^2 , \ldots , g_n(z) = \frac{1}{2} |z_n|^2 ,$$

say $g_i = a_i$, none of the a_i's being zero, so that L is a Lagrangian torus. Choose a subtorus of T^n:

$$T^{n-1} = \{(t_1, \ldots, t_n) \in T^n \mid t_1 \cdots t_n = 1\} .$$

Let V be an orbit of this subtorus, an isotropic torus of dimension $n - 1$. Consider the Hamiltonian vector fields $Y_1, \ldots, Y_n$ associated to the functions g_i:

$$\begin{cases} Y_1(z_1, \ldots, z_n) = (iz_1, 0, \ldots, 0) \\ \vdots \\ Y_n(z_1, \ldots, z_n) = (0, \ldots, 0, iz_n). \end{cases}$$

Let $z = (z_1, \ldots, z_n)$ be a point of V. The tangent space to L at z is spanned by the values of the Y_i's, the tangent space to V is the hyperplane consisting of the vectors $\sum \lambda_i Y_i$ satisfying $\sum \lambda_i = 0$. It is spanned by the values at z of the vector fields

$$X_1 = Y_1 - Y_n, \ldots, X_{n-1} = Y_{n-1} - Y_n,$$

that are the Hamiltonian vector fields of the functions

$$f_1 = g_1 - g_n, \ldots, f_{n-1} = g_{n-1} - g_n.$$

We are looking now for an n-th function f such that the subspace spanned by the vectors $X_1, \ldots, X_{n-1}$ and X_f is a special Lagrangian at each point where the vectors are independent. The subspace $F = \langle X_1, \ldots, X_{n-1} \rangle$ is isotropic and has dimension $n-1$.

We look for X_f as a linear combination $X_f = \sum \lambda_j Y_j$ such that:

- The vector field X_f is in the subspace orthogonal to $\langle X_1, \ldots, X_{n-1} \rangle$ (for the Hermitian form), that is, $\langle X_f, X_k \rangle = 0$ for $1 \le k \le n-1$. This gives $\lambda_k |z_k|^2 - \lambda_n |z_n|^2 = 0$. Thus λ_k must have the form

$$\lambda_k = \frac{\mu(z_1, \ldots, z_n)}{|z_k|^2}.$$

- The determinant

$$\begin{vmatrix} iz_1 & & 0 & \lambda_1 iz_1 \\ 0 & \ddots & 0 & \vdots \\ \vdots & & iz_{n-1} & \vdots \\ -iz_n & \cdots & -iz_n & \lambda_n iz_n \end{vmatrix}$$

is real. This allows to determine the function μ.

Subtracting the linear combination $\lambda_1 X_1 + \cdots + \lambda_{n-1} X_{n-1}$ from the last vector, this vector becomes $(\lambda_1 + \cdots + \lambda_n) Y_n$, so that the determinant is $i^n (\lambda_1 + \cdots + \lambda_n) z_1 \cdots z_n$. We are thus looking for functions f and μ such that

$$X_f(z_1, \ldots, z_n) = (\lambda_1 iz_1, \ldots, \lambda_n iz_n) \quad \text{and} \quad i^n (\mu(z_1, \ldots, z_n)) z_1 \cdots z_n \text{ is real.}$$

For any index j, we must have:

$$2 \frac{\partial f}{\partial z_j} = \bar{\lambda}_j \bar{z}_j, \quad 2 \frac{\partial f}{\partial \bar{z}_j} = \lambda_j z_j, \quad \text{and } i^n \mu(z_1, \ldots, z_n) z_1 \cdots z_n \in \mathbf{R}.$$

The functions

$$f(z_1, \ldots, z_n) = z_1 \cdots z_n + \overline{z_1 \cdots z_n}, \quad \mu = 2 \bar{z}_1 \cdots \bar{z}_n$$

give a solution when $i^n \in \mathbf{R}$, namely when n is even. When n is odd, we rather take

$$f(z_1, \ldots, z_n) = \frac{1}{i}(z_1 \cdots z_n - \overline{z_1 \cdots z_n}), \quad \mu = 2 \bar{z}_1 \cdots \bar{z}_n$$

Proposition I.5.17. *The functions* $f_1, \ldots, f_n$ *defined by*

$$f_1(z_1, \ldots, z_n) = \frac{1}{2}(|z_1|^2 - |z_n|^2), \ldots, f_{n-1}(z_1, \ldots, z_n) = \frac{1}{2}(|z_{n-1}|^2 - |z_n|^2)$$

and

$$f_n(z_1, \ldots, z_n) = \begin{cases} \text{Re}(z_1 \cdots z_n) & \text{if } n \text{ is even} \\ \text{Im}(z_1 \cdots z_n) & \text{if } n \text{ is odd} \end{cases}$$

form an integrable system on $\mathbf{C}^n = \mathbf{R}^n \times \mathbf{R}^n$, *all the regular common level sets of which are special Lagrangian cylinders* $T^{n-1} \times \mathbf{R}$.

Proof. The only thing that is left to prove is that the regular levels are "cylinders" $T^{n-1} \times \mathbf{R}$. As we are dealing with an integrable system, we know that the levels are endowed with an $\mathbf{R}^n$-action. Here the $n-1$ first vector fields are periodic and in particular complete; the last one is complete too, because the level is a closed submanifold of $\mathbf{C}^n$. The action is thus an action of $T^{n-1} \times \mathbf{R}$ and this is a free action, as the level, being special Lagrangian, cannot be compact. □

Exercise I.19 describes essentially the same construction.

I.5.f   Special Lagrangian submanifolds invariant under $\mathrm{SO}(n)$

The next and sporadic examples also come from [18]. Start from a smooth curve Γ in

$$\mathbf{C} = \mathbf{C} \times \{0\} \subset \mathbf{C} \times \mathbf{C}^{n-1} = \mathbf{C}^n$$

and "rotate" it with the help of the diagonal $\mathrm{SO}(n)$-action, namely

$$g \cdot (X + iY) = g \cdot X + ig \cdot Y \text{ for } g \in \mathrm{SO}(n) \text{ and } X, Y \in \mathbf{R}^n.$$

If we assume the curve does not pass through 0, we get a submanifold of $\mathbf{C}^n$:

$$V = \{(x + iy)u \mid x + iy \in \Gamma, u \in \mathbf{R}^n, u = g(e_1) \text{ for some } g \in \mathrm{SO}(n)\}$$

(notice that u describes a sphere $S^{n-1} \subset \mathbf{R}^n$). The tangent space to V at $(x+iy)u$ is spanned by the vectors $(x+iy)U$ with $U \in T_u S^{n-1}$ and the $(\xi+i\eta)u$ with $\xi+i\eta$ tangent to Γ at $x + iy$. The submanifold V is always Lagrangian, as is easily checked:

$$\omega((x + iy)U, (x + iy)U') = xy(U \cdot U' - U' \cdot U) = 0,$$
$$\omega((x + iy)U, (\xi + i\eta)u) = (x\eta - y\xi)U \cdot u = 0.$$

It is special Lagrangian if and only if, denoting $(U_1, \ldots, U_{n-1})$ a basis of $T_u S^{n-1}$,

$$\det_{\mathbf{C}} ((x + iy)U_1, \ldots, (x + iy)U_{n-1}, (\xi + i\eta)u) \in \mathbf{R}.$$

But this determinant is equal to $(x + iy)^{n-1}(\xi + i\eta) \det_{\mathbf{C}}(U_1, \ldots, U_{n-1}, u)$, or to $(x+iy)^{n-1}(\xi+i\eta) \det_{\mathbf{R}}(U_1, \ldots, U_{n-1}, u)$ since these vectors are in $\mathbf{R}^n \subset \mathbf{C}^n$. The condition is thus that

$$(x + iy)^{n-1}(\xi + i\eta) \in \mathbf{R} \text{ for any tangent vector } \xi + i\eta \text{ to } \Gamma.$$

We get eventually:

Proposition I.5.18. *The Lagrangian submanifold of* $\mathbf{C}^n$

$$V = \{(x + iy)u \mid (x + iy) \in \Gamma, u \in S^{n-1} \subset \mathbf{R}^n\}$$

is special Lagrangian if and only if, on Γ, *the function* $\mathrm{Im}\,((x + iy)^n)$ *is constant.* □

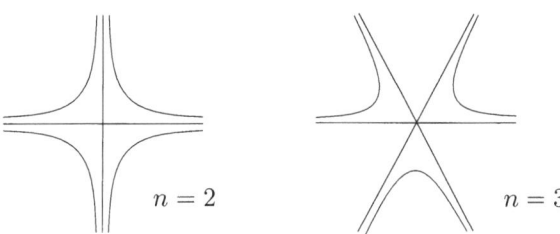

Figure I.9

Remark I.5.19. This method gives essentially one special Lagrangian submanifold in any dimension, which is not much!

Remark I.5.20. Any connected component of Γ is diffeomorphic to $\mathbf{R}$, the special Lagrangian submanifolds obtained are (unions of) copies of $S^{n-1} \times \mathbf{R}$.

To draw a picture of the special Lagrangian submanifold, one draws first the curve Γ (in the (x, y) plane), then its wave front (in the (x, z) plane). One then notices that the Liouville form $\lambda = Y \cdot dX$ is, on V:

$$\lambda = Y \cdot dX = (yu) \cdot d(xu)$$
$$= (yu) \cdot ((dx)u + xdu)$$
$$= ydx$$

(since $udu = \frac{1}{2}\|u\|^2 = 0$) so that the wave front of V is

$$\{(xu, z) \in \mathbf{R}^n \times \mathbf{R} \mid (x, z) \text{ is a point of the wave front of } \Gamma\}.$$

For example, for $n = 2$, the curve Γ is a hyperbola $xy = $ constant, its wave front is the curve $z = \log x$ and the wave front of the special Lagrangian submanifold is the surface of revolution obtained by rotating the graph of the logarithm function about the z-axis (Figure I.10).

I.6   Appendices

I.6.a   The topology of the symplectic group

Proposition I.6.1. *The manifold* $\mathrm{Sp}(2n)$ *is diffeomorphic to the Cartesian product of the group* $\mathrm{U}(n)$ *with a convex open cone of a vector space of dimension* $n(n+1)$.

Corollary I.6.2. *The symplectic group* $\mathrm{Sp}(2n)$ *is path connected. The injection of* $\mathrm{U}(n)$ *in* $\mathrm{Sp}(2n)$ *induces an isomorphism*

$$\mathbf{Z} = \pi_1 \, \mathrm{U}(n) \longrightarrow \pi_1 \, \mathrm{Sp}(2n).$$

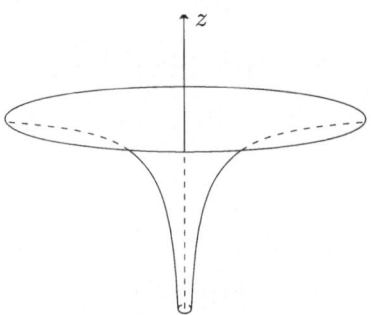

Figure I.10

Proof of the proposition. Let $A \in \mathrm{Sp}(2n)$. As any invertible transformation of $\mathbf{R}^{2n}$, A can be written in a unique way as a product

$$A = S \cdot \Omega$$

where S is the positive definite symmetric matrix $S = \sqrt{A^t A}$ and Ω is the orthogonal matrix $\Omega = S^{-1}A$. As A is symplectic, the matrix S is also symplectic: $^t A$ and $A^t A$ are symplectic, the matrix $A^t A$ is symmetric, positive definite, thus it is diagonalizable in an orthonormal basis and S is the matrix that, in this basis, is the diagonal of the square roots of the eigenvalues of $A^t A$, so that S is indeed symplectic as is $A^t A$. One deduces that

$$\Omega = S^{-1}A \in \mathrm{Sp}(2n) \cap \mathrm{O}(2n) = \mathrm{U}(n)$$

and thus that Ω is a unitary matrix. We have thus obtained a bijection

$$\mathrm{Sp}(2n) \longrightarrow \mathrm{U}(n) \times \mathcal{S}$$
$$A \longmapsto ((\sqrt{A^t A})^{-1}A, \sqrt{A^t A})$$

where $\mathcal{S}$ denotes the set of positive definite symmetric matrices that are symplectic. We still have to prove that this space is an open convex cone in a vector space of dimension $n(n+1)$. Write the matrices as block matrices in a symplectic basis. Let $S \in \mathcal{S}$, we have

$$S = \begin{pmatrix} A & B \\ {}^t B & C \end{pmatrix} \text{ with } A \text{ and } C \text{ positive definite symmetric and } {}^t SJS = J.$$

The last condition, that expresses the fact that S is symplectic, is equivalent to

$$BA \text{ is symmetric and } C = A^{-1}(\mathrm{Id} + B^2).$$

The mapping

$$\mathcal{S} \longrightarrow \mathrm{Sym}(n; \mathbf{R}) \times \mathrm{Sym}^+(n; \mathbf{R})$$
$$S \longmapsto \qquad (BA, A)$$

is the desired diffeomorphism. The open set $\mathrm{Sym}^+(n; \mathbf{R})$ of all positive definite symmetric real matrices is obviously an open convex cone in the vector space $\mathrm{Sym}(n; \mathbf{R})$ of all symmetric matrices, the product is an open convex cone of the product space, that has dimension $2\dfrac{n(n+1)}{2}$. □

Proof of the corollary. The convex cone $\mathrm{Sym}(n; \mathbf{R}) \times \mathrm{Sym}^+(n; \mathbf{R})$ is contractible.
□

Remark I.6.3. There is another beautiful proof of this type of contractibility results, due to Sévennec, in [5].

I.6.b Complex structures

If E is a vector space endowed with a symplectic from ω, it is said that an endomorphism J of E is a complex structure *calibrated* by ω if $J^2 = -\mathrm{Id}$ (J is a complex structure),

$$\omega(Jv, Jw) = \omega(v, w)$$

(J is symplectic) and

$$g(v, w) = \omega(v, Jw)$$

is a scalar product (namely a positive definite bilinear form) on E.

I.6.c Hamiltonian vector fields, integrable systems

In this appendix, denote for simplicity $\mathbf{C}^n = \mathbf{R}^{2n}$ by W. It can be replaced by any symplectic manifold W (see § II.1).

Hamiltonian vector fields

To any function $H : W \to \mathbf{R}$, the symplectic form allows to associate a vector field, a kind of gradient, the *Hamiltonian vector field* X_H (sometimes called the "symplectic gradient" H). This is the vector field defined by the relation

$$\omega_x(Y, X_H(x)) = (dH)_x(Y) \text{ for all } Y \in T_x W,$$

or by

$$\iota_{X_H}\omega = -dH.$$

In coordinates, one has

$$X_H(x_1, \ldots, x_n, y_1, \ldots, y_n) = \left(\frac{\partial H}{\partial y_1}, \ldots, \frac{\partial H}{\partial y_n}, -\frac{\partial H}{\partial x_1}, \ldots, -\frac{\partial H}{\partial x_n}\right).$$

Notice that the vector field X_H vanishes at x if and only if x is a critical point of the function H:

$$X_H(x) = 0 \iff (dH)_x = 0.$$

In particular, the singularities (or zeroes) of a Hamiltonian vector field are the critical points of a function.

Notice also that the function H is constant along the trajectories, or integral curves, of the vector field X_H: as ω_x is skew symmetric, we have $(dH)(X_H) = 0$ or $X_H \cdot H = 0$.

The Poisson bracket

Assume now that f and g are two functions on W. Define their "Poisson bracket" $\{f, g\}$ by the formula

$$\{f, g\} = X_f \cdot g = dg(X_f).$$

In coordinates, one has

$$\{f, g\} = \sum_{i=1}^{n} \left(\frac{\partial f}{\partial y_i} \frac{\partial g}{\partial x_i} - \frac{\partial g}{\partial y_i} \frac{\partial f}{\partial x_i} \right).$$

Notice that

$$X_f \cdot g = dg(X_f) = \omega(X_f, X_g) = -\omega(X_g, X_f) = -df(X_g) = -X_g \cdot f,$$

so that $\{f, g\} = -\{g, f\}$. This shows that the Poisson bracket is skew-symmetric in f and g. By definition, this is also a derivation (in both entries); in other words, the Poisson bracket satisfies the Leibniz identity

$$\{f, gh\} = \{f, g\} h + g \{f, h\}.$$

Using the general relation

$$\mathcal{L}_X \iota_Y - \iota_Y \mathcal{L}_X = \iota_{[X,Y]}$$

and Cartan formula

$$\mathcal{L}_X = d\iota_X + \iota_X d,$$

we get

$$\iota_{[X_f, X_g]} \omega = \mathcal{L}_{X_f} \iota_{X_g} \omega - \iota_{X_g} \mathcal{L}_{X_f} \omega$$
$$= d\iota_{X_f} \iota_{X_g} \omega + \iota_{X_f} d\iota_{X_g} \omega - \iota_{X_g} d\iota_{X_f} \omega - \iota_{X_g} \iota_{X_f} d\omega$$
$$= d\iota_{X_f} \iota_{X_g} \omega = d(\omega(X_g, X_f)) = -d \{f, g\},$$

in other words

$$[X_f, X_g] = X_{\{f, g\}}.$$

We also have

$$[X_f, X_g] \cdot h = \{\{f, g\}, h\}.$$

From this, we deduce that the Poisson bracket satisfies the Jacobi identity

$$\{f, \{g, h\}\} + \{g, \{h, f\}\} + \{h, \{f, g\}\} = 0$$

and thus defines a Lie algebra structure on $\mathcal{C}^\infty(W)$, the mapping

$$\mathcal{C}^\infty(W) \longrightarrow \mathfrak{X}(W)$$
$$f \longmapsto X_f$$

being a morphism of Lie algebras from $\mathcal{C}^\infty(W)$ (with the Poisson bracket) into the Lie algebra of vector fields (with the Lie bracket of vector fields).

Proof of the Jacobi identity. Apply the definition of the bracket of vector fields:

$$[X_f, X_g] \cdot h = X_f \cdot (X_g \cdot h) - X_g \cdot (X_f \cdot h),$$

and the equality above to get

$$\begin{aligned}
\{\{f, g\}, h\} &= [X_f, X_g] \cdot h \\
&= X_f \cdot (X_g \cdot h) - X_g \cdot (X_f \cdot h) \\
&= X_f \cdot \{g, h\} - X_g \cdot \{f, h\} \\
&= \{f, \{g, h\}\} - \{g, \{f, h\}\}.
\end{aligned}$$

This, taking into account the skew-symmetry of the Poisson bracket, is equivalent to the Jacobi identity.   □

Integrable systems

As any vector field does it, the Hamiltonian vector field X_H defines a differential system on W, namely,

$$\dot{x}(t) = X_H(x(t)),$$

the *Hamiltonian system* associated with H. The function H is constant along the trajectories of this system, in other words

$$X_H \cdot H = 0 \text{ or } dH(X_H) = 0.$$

It is said that H is a first integral of the system. More generally, a function $f : W \to \mathbf{R}$ that is constant along the integral curves of a vector field X is called a *first integral* of X. In the case of a Hamiltonian vector field X_H, the equality $X_H \cdot f = 0$ is equivalent to $\{f, H\} = 0$, we say that the functions f and H *commute*.

It is said that a Hamiltonian system is integrable if it has "as many commuting first integrals as possible". Let us explain this:

- Let $f_1, \ldots, f_k$ be commuting first integrals of the system X_H, so that $\{f_i, f_j\} = 0$ for all i and j. Each one is constant on the trajectories of the Hamiltonian system associated to each other one.

- The expression "as many as possible": at any point x of W, the subspace of $T_x W$ spanned by the Hamiltonian vector fields of the functions f_i is isotropic:

$$\omega(X_{f_i}, X_{f_j}) = \pm\{f_i, f_j\} = 0.$$

  Its dimension is thus at most $n = \frac{1}{2}\dim W$. It is required that, at least for x in an open dense subset of W, this subspace has maximal dimension n.

- Notice that the vectors X_{f_i} are independent at x if and only if the linear forms $(df_i)_x$ are independent.

Definition I.6.4. The function H or the Hamiltonian vector field X_H on W is qualified as *integrable* if it has n independent commuting first integrals.

Examples I.6.5. Every function depending only of the coordinates y_i,

$$H = H(y_1, \ldots, y_n)$$

is integrable: the functions y_i are independent commuting first integrals. Every Hamiltonian system on $\mathbf{C}$ is integrable. Similarly, a Hamiltonian system on $\mathbf{C}^2$ is integrable if and only if it has a "second first integral".

Exercises

Exercise I.1. Let V be a real vector space and V^* be its dual. Check that the form ω defined on $V \oplus V^*$ by

$$\omega((v, \alpha), (w, \beta)) = \alpha(w) - \beta(v)$$

is a symplectic form

Exercise I.2 *(Relative linear Darboux theorem).* Let F be a vector subspace of a symplectic vector space E. Assume that the restriction of the symplectic form to F has rank $2r$. Show that there exists a symplectic basis $(e_1, \ldots, e_n, f_1, \ldots, f_n)$ of E such that $(e_1, \ldots, e_r, e_{r+1}, \ldots, e_{r+k}, f_1, \ldots, f_r)$ is a basis of F (k is the integer defined by $2r + k = \dim F$).

Exercise I.3. Show that the symplectic group of $\mathbf{C}$ is isomorphic with the special linear group $\mathrm{SL}(2; \mathbf{R})$.

Exercise I.4. Prove directly that the symplectic group $\mathrm{Sp}(2)$ is diffeomorphic to the product of a circle by an open disk.

Exercise I.5. Let $A \in \mathrm{Sp}(2n)$. Check that the matrices ${}^t A$ and A^{-1} are similar[8]. Show that λ is an eigenvalue of A if and only if λ^{-1} is also an eigenvalue, and that both occur with the same multiplicity.

---

[8]Thus A and A^{-1} are similar too.

Exercise I.6. Check that a non zero vector of a symplectic space can be mapped to any other non zero vector by a symplectic transformation (in other words, the symplectic groups acts transitively on the set of non zero vectors).

Show that, for $n > 1$, the symplectic group does not act transitively on the set of real 2-dimensional subspaces of $\mathbf{C}^n$.

Exercise I.7. Let $n > 1$ be an integer. Let P be a real plane (dimension-2 subspace) in $\mathbf{C}^n$. Show that P is either isotropic or symplectic. What are the orbits of the action of the symplectic group on the set of planes in $\mathbf{C}^n$?

Exercise I.8. Let V be a vector space and $V^\star$ be its dual. Endow $V \oplus V^\star$ with the symplectic form defined in Exercise I.1. Let $A : V \to V^\star$ be a linear map. Prove that the graph of A is a Lagrangian subspace if and only if the bilinear form defined by A on V is symmetric.

Exercise I.9. Let E be a vector space endowed with a symplectic form ω and let F be (any) subspace of E. Prove that ω induces a symplectic structure on the quotient $F/F \cap F^\circ$.

Exercise I.10. Let E be an even dimensional vector space and let ω, ω' be two symplectic forms on E. Prove that the symplectic groups $\mathrm{Sp}(E, \omega)$ and $\mathrm{Sp}(E, \omega')$ are conjugated subgroups of $\mathrm{GL}(E)$.

Let $\Omega(E)$ be the space of all symplectic forms on the vector space E. Prove that the linear group of E acts on this space by

$$(g \cdot \omega)(X, Y) = \omega(gX, gY).$$

Deduce that $\Omega(E)$ is in one-to-one correspondence[9] with the homogeneous space $\mathrm{GL}(E)/\mathrm{Sp}(E)$, where $\mathrm{Sp}(E)$ is the symplectic group $\mathrm{Sp}(E, \omega_0)$ for a given form ω_0 on E.

Exercise I.11. Prove that, on any symplectic vector space, there are complex structures. Prove that a complex structure is an isometry and that it is skew-symmetric for the scalar product it defines.

Exercise I.12. Let V be a real vector space. Using a scalar product on V, construct a complex structure calibrated by the standard symplectic form on $V \oplus V^\star$ and such that

$$(J(v), w) = v \cdot w \text{ for all } v, w \in V.$$

Exercise I.13. Assume that the wave front

$$] - \alpha, \alpha[\longrightarrow \mathbf{R}^2$$
$$t \longmapsto (x(t), z(t))$$

has an ordinary cusp for $t = 0$ with a tangent line transversal to the z-axis. Prove that this is the wave front of a Lagrangian *immersion* of $] - \alpha, \alpha[$ into $\mathbf{R}^2$.

---

[9]This is actually a homeomorphism.

Exercise I.14. Prove that the wave front of the Whitney immersion $S^n \to \mathbf{C}^n$ is the hypersurface in $\mathbf{R}^{n+1}$ image of the sphere S^n by

$$(x, a) \longmapsto \left(x, a\, \|x\|^2 + \frac{a^3}{3} - a \right)$$

(using the notation of Example I.4.3). Find the singular points of this wave front and draw it in the cases $n = 1$ (this is the eye, Figure I.2) and $n = 2$ (this is the flying saucer, Figure I.4).

Exercise I.15 *(The swallow tail).* Determine... and draw the wave front of the Lagrangian immersion described in §I.4.5 and on Figure I.1.

Exercise I.16. Prove that the Maslov class of the standard (Lagrangian) embedding of the circle is ± 2. What is that of the Whitney immersion? Of the immersion defined by the crossbow[10]?

Exercise I.17 *(Lagrangian cobordisms* [3]*).* The space $\mathbf{C}^n$ is endowed with its Liouville form λ and its symplectic form $d\lambda$. It is said that a Lagrangian immersion $f : L \to \mathbf{C}^n$ is "cobordant to zero" if there exists an *oriented* manifold V of dimension $n + 1$, with boundary, whose boundary is L, and a Lagrangian immersion

$$\widetilde{f} : V \longrightarrow \mathbf{C}^{n+1}$$

transversal to the co-isotropic subspace $F = \mathbf{C}^n \oplus i\mathbf{R} \subset \mathbf{C}^{n+1}$, such that

$$\widetilde{f}^{-1}(F \cap V) = \partial V = L$$

and such that the composition

$$L \xrightarrow{\;\widetilde{f}|_L\;} F \longrightarrow F/F^\circ = \mathbf{C}^n$$

is the immersion f.

(1) Prove that the Whitney immersion $S^n \to \mathbf{C}^n$ (§I.4.3) is cobordant to zero.

(2) Assume that $f : S^1 \to \mathbf{C}$ is cobordant to zero. What can be said of $\int_{S^1} f^\star \lambda$? Prove that, if a Lagrangian immersion $S^1 \to \mathbf{C}$ is cobordant to zero, it is exact.

(3) Consider an exact Lagrangian immersion

$$f : S^1 \longrightarrow \mathbf{C}$$

and its wave front in $\mathbf{R}^2$. Assume the singularities of the wave front are ordinary cusps. The tangent line to the front at any point is transversal

---

[10]Hint: orient the circle and notice that the unit tangent vector to the Whitney immersion does not take all the values in the circle. For the crossbow, notice that this immersion of the circle into $\mathbf{C}$ may be deformed, among immersion, into the standard embedding.

to the z-axis. The circle S^1 is oriented. Count the cusps of type (a) with a + sign, those of type (b) with a − sign (Figure I.11) and get a number $N(f) \in \mathbf{Z}$. What is the value of $N(f)$ for the Whitney immersion? For the crossbow (Figure I.3)?

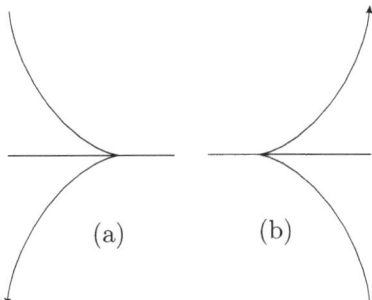

(a)   (b)

Figure I.11

(4) The Lagrangian immersion $f : S^1 \to \mathbf{C}$ has a Gauss map $\gamma(f)$, taking its values in the Grassmannian $\tilde{\Lambda}_1$ of oriented Lagrangians in $\mathbf{C}$, that is a circle S^1. Call σ the closed 1-form "$d\theta$" on this circle. Prove that[11]

$$N(f) = \frac{1}{2\pi} \int_{S^1} \gamma(f)^\star \sigma.$$

(5) Consider the mapping

$$j : \tilde{\Lambda}_1 \longrightarrow \tilde{\Lambda}_2$$
$$P \longmapsto P \oplus \mathbf{R} \subset \mathbf{C} \oplus \mathbf{R} \subset \mathbf{C}^2.$$

It can be shown (this is an additional question, use § I.3.f) that

$$j^\star : H^1(\tilde{\Lambda}_2) \longrightarrow H^1(\tilde{\Lambda}_1)$$

is an isomorphism. Prove that if $f : S^1 \to \mathbf{C}$ is cobordant to zero, then $N(f) = 0$. Does there exist a Lagrangian immersion of a disk into $\mathbf{C}^2$ whose boundary is the crossbow?

---

[11] This is to say that $N(f)$ is the Maslov class of the immersion f.

Exercise I.18 *(From (x,y) to $(z, \bar{z})$)*. Writing

$$dz = dx + i dy, \quad d\bar{z} = dx - i dy$$

one gets a couple of relations between the expressions of the vector fields in coordinates (x, y) or $(z, \bar{z})$. Prove for instance that

$$X_f = \sum_{j=1}^n \left(\frac{\partial f}{\partial y_j} \frac{\partial}{\partial x_j} - \frac{\partial f}{\partial x_j} \frac{\partial}{\partial y_j} \right)$$

$$= \frac{i}{2} \sum_{j=1}^n \left(\frac{\partial f}{\partial z_j} \frac{\partial}{\partial \bar{z}_j} - \frac{\partial f}{\partial \bar{z}_j} \frac{\partial}{\partial z_j} \right).$$

Exercise I.19. Consider the vector field X given on $\mathbf{C}^2$ by

$$X(z_1, z_2) = (i\alpha_1 z_1, i\alpha_2 z_2)$$

(α_1 and α_2 being two real parameters).

(1) Check that

$$X(z_1, z_2) = \alpha_1 \left(iz_1 \frac{\partial}{\partial z_1} - i\bar{z}_1 \frac{\partial}{\partial \bar{z}_1} \right) + \alpha_2 \left(iz_2 \frac{\partial}{\partial z_2} - i\bar{z}_2 \frac{\partial}{\partial \bar{z}_2} \right)$$

and show that the form $\iota_X \Omega$ is holomorphic.

(2) Show that X preserves ω and find a function H such that $X = X_H$.

(3) Under which condition does the vector field X preserve Ω? Assume now that this condition holds. Find two functions g and h from $\mathbf{C}^2$ to $\mathbf{R}$ such that

$$\iota_X \Omega = dg + i dh.$$

Consider $H^{-1}(a) \cap h^{-1}(b)$. Show that, if a is a regular value of H, this is a special Lagrangian submanifold.

(4) Describe the special Lagrangian submanifolds $H^{-1}(a) \cap h^{-1}(b)$ as complex j-curves, that is, by equations.

(5) Check that they are diffeomorphic to $S^1 \times \mathbf{R}$. Hint: they are conics.

Chapter II

Lagrangian and special Lagrangian submanifolds in symplectic and Calabi-Yau manifolds

II.1 Symplectic manifolds

In order to deform a Lagrangian submanifold in $\mathbf{C}^n$, we must understand how a tubular neighbourhood looks like. We prove here that a Lagrangian submanifold has a neighbourhood which is diffeomorphic to a neighbourhood of the zero section in its cotangent bundle. To be precise and explicit, we need to define a symplectic structure on the cotangent bundles and more generally to say what a symplectic structure on a manifold is.

A *symplectic manifold* is a manifold W endowed with a non degenerate 2-form ω, namely, a non degenerate alternated bilinear form ω_x on each tangent space $T_x W$, which is required to be *closed*, $(d\omega = 0)$. Notice that a symplectic manifold is even dimensional.

Examples II.1.1.   (1) The first example is of course $\mathbf{C}^n$ with the symplectic form we have used so far, considered as a differential form:

$$\omega = \sum_{j=1}^{n} dy_j \wedge dx_j$$

(where $(x_1 + iy_1, \ldots, x_n + iy_n)$ stands for the complex coordinates in $\mathbf{C}^n$).

One also has:

$$\omega_z(Z, Z') = \omega(Z, Z') = \sum_{j=1}^{n}(X'_j Y_j - X_j Y'_j) = X' \cdot Y - X \cdot Y'.$$

And this is an exact, hence closed, form:

$$\omega = d\Big(\sum_{j=1}^{n} y_j dx_j\Big).$$

(2) The next example is that of cotangent bundles. Think that $\mathbf{C}^n = \mathbf{R}^n \times \mathbf{R}^n$, then that $\mathbf{R}^n \times \mathbf{R}^n = T^*\mathbf{R}^n$ and simply replace $\mathbf{R}^n$ by any manifold V. On $W = T^*V$, there is a *canonical* 1-form, the *Liouville form* λ, defined by the "compact" formula:

$$\lambda_{(x,\alpha)}(X) = \alpha\left(T_{(x,\alpha)}\pi(X)\right)$$

... in which x denotes a point of V, α an element of T_x^*V (namely a linear form on the tangent space T_xV), and π the projection $T^*V \to V$ of the cotangent bundle. If $(x_1, \ldots, x_n)$ are local coordinates on V and $(y_1, \ldots, y_n)$ the cotangent coordinates, then

$$\lambda = \sum_{j=1}^{n} y_j dx_j.$$

The 2-form $d\lambda$ is both closed (!) and non degenerate.

(3) Surfaces. On a surface W, any 2-form is closed. Moreover, in dimension 2, to say that a 2-form is non degenerate means that it nowhere vanishes, in other words that this is a volume form: all the orientable surfaces may be considered as symplectic manifolds.

(4) The sphere. Consider, in particular, the unit sphere S^2 in $\mathbf{R}^3$, whose tangent space at a point v is the plane orthogonal to the unit vector v. Put

$$\omega_v(X, Y) = v \cdot (X \wedge Y) = \det(v, X, Y).$$

This is a non degenerate 2-form and thus a symplectic form.

(5) The projective space $\mathbf{P}^n(\mathbf{C})$ is a symplectic manifold. The nicest thing to do is to define its symplectic form starting from that of $\mathbf{C}^{n+1}$ and using the symplectic reduction process. To define $\mathbf{P}^n(\mathbf{C})$, we factor out the unit sphere S^{2n+1} of $\mathbf{C}^{n+1}$ by the S^1-action (multiplication of coordinates):

$$t \cdot (z_1, \ldots, z_{n+1}) = (tz_1, \ldots, tz_{n+1})$$

At each point x of the sphere S^{2n+1}, the tangent space is the Euclidean orthogonal of x and the kernel of the restriction of the symplectic from is the line generated by ix. This line is also the tangent space to the circle through x on the sphere.

The symplectic form of $\mathbf{C}^{n+1}$ defines a non degenerate alternated bilinear form ω on $\mathbf{P}^n(\mathbf{C})$. Its pull-back on the sphere is closed, so that ω is closed. It is actually a (the standard) Kähler form on $\mathbf{P}^n(\mathbf{C})$.

(6) *Complex* submanifolds of the projective space are symplectic. The compatibility of ω with the complex structure gives that $\omega(X, iX) > 0$ for any vector X that is tangent to the submanifold, so that ω is indeed non degenerate on this submanifold.

(7) More generally, all Kähler manifolds are symplectic. We will come back to this remark.

Notice that, on cotangent bundles, as on $\mathbf{C}^n$, the symplectic form is *exact*. This cannot be the case on a compact symplectic manifold.

Proposition II.1.2. *On a compact manifold, there exists no 2-form that is both non degenerate and exact.*

Proof. Let ω be a non degenerate 2-form on the $2n$-dimensional manifold W. To say that ω is non degenerate is to say that $\omega^{\wedge n}$ is a *volume form*. But then, if $\omega = d\alpha$,

$$\omega^{\wedge n} = d(\alpha \wedge \omega^{\wedge(n-1)})$$

is also exact, thus W cannot be compact. $\qquad\square$

Hamiltonian vector fields X_H for functions $H : W \to \mathbf{R}$ are defined exactly as in Appendix I.6.c and so is the Poisson bracket of two functions on W. Exercise II.3 explains why it is required that a symplectic form be closed.

II.2 Lagrangian submanifolds and immersions

An immersion $f : L \to W$ into a symplectic manifold is *Lagrangian* if $f^\star\omega = 0$ and $\dim W = 2 \dim L$.

II.2.a In cotangent bundles

All what was done in $\mathbf{C}^n$ in §I.4.a works as well in a cotangent bundle.

Graphs

Proposition I.4.1 generalizes as:

Proposition II.2.1. *Let $\alpha : L \to T^*L$ be a section of a cotangent bundle. Its image is a Lagrangian submanifold if and only if the 1-form α is closed.*

Proof. The most elegant thing to do is to state first a property of the Liouville form (which explains why it is called the "canonical" 1-form): for any form α, one has

$$\alpha^\star \lambda = \alpha.$$

In this equality, α is considered as a section of the cotangent bundle in the left hand side and as a form in the right hand side. One has indeed:

$$\begin{aligned}
(\alpha^\star \lambda)_x(Y) &= \lambda_{(x,\alpha_x)}(T_x\alpha(Y)) && \text{by definition of } \alpha^\star \\
&= \alpha_x(T_{(x,\alpha_x)}\pi \circ T_x\alpha(Y)) && \text{by definition of } \lambda \\
&= \alpha_x(Y) && \text{because } \alpha \text{ is a section.}
\end{aligned}$$

Eventually, $\alpha^\star \omega = 0$ if and only if $d(\alpha^\star \lambda) = 0$, thus the graph of α is a Lagrangian submanifold if and only if α is closed. $\qquad\square$

Remark II.2.2. In particular, the zero section of $L \subset T^*L$ is a Lagrangian submanifold. What we plan to do next is to show that $L \subset T^*L$ is a model for *all* Lagrangian embeddings of L into a symplectic manifold (Theorem II.3.7).

Generating functions

A function

$$F : M \times \mathbf{R}^k \longrightarrow \mathbf{R}$$

allows to construct a Lagrangian submanifold (the graph of dF) into $T^*M \times \mathbf{C}^k$ and then, by reduction, a Lagrangian immersion into T^*M.

Wave fronts

Exact Lagrangian immersions into T^*M define wave fronts in $M \times \mathbf{R}$ and conversely.

Conormal bundles

Let

$$f : V \longrightarrow M$$

be any immersion. The *conormal* bundle is the subbundle of the pull back bundle

$$f^*T^*M = \left\{ (x, \varphi) \mid x \in V, \varphi \in T^*_{f(x)}M \right\} \longrightarrow V$$

defined by

$$N^\star f = \left\{ (x, \varphi) \in f^\star T^\star M \mid \varphi|_{T_x f(T_x V)} = 0 \right\}$$
$$= \left\{ (x, \varphi) \in f^\star T^\star M \mid \varphi \circ T_x f = 0 \right\}.$$

Map $N^\star f$ into $T^\star M$ by

$$F : (x, \varphi) \longmapsto (f(x), \varphi).$$

This is an immersion, since

$$T_{(x,\varphi)} F(\xi, \psi) = (T_x f(\xi), \psi).$$

It is Lagrangian, as we have $F^\star \lambda = 0$. Indeed, calling π the two projections $T^\star M \to M$ et $N^\star f \to V$, we get

$$(F^\star \lambda)_{(x,\varphi)}(X) = \lambda_{(f(x),\varphi)} \left(T_{(x,\varphi)} F(X) \right)$$
$$= \varphi \left(T_{(f(x),\varphi)} \pi \circ T_{(x,\varphi)} F(X) \right)$$
$$= \varphi \left(T_x f(T_{(x,\varphi)} \pi(X)) \right)$$
$$= 0$$

as φ vanishes on the vectors that are tangent to V. $\qquad\square$

One should check that the proof given for the normal bundle in §I.4.c for the case where $M = \mathbf{R}^n$ is identical to the one given here, the orthogonality used there being an *ersatz* of the duality used here.

II.3 Tubular neighborhoods of Lagrangian submanifolds

Let us now present a method, invented by Moser [28], which allows to describe a symplectic manifold in the neighbourhood of a point (they are all the same) or a neighbourhood of a Lagrangian submanifold in a symplectic manifold.

II.3.a Moser's method

The next "lemma" contains all these results.

Lemma II.3.1. *Let W be a $2n$-dimensional manifold and let $Q \subset W$ be a compact submanifold. Assume that ω_0 and ω_1 are two closed 2-forms on W such that, at any point x of Q, ω_0 and ω_1 are equal and non degenerate on $T_x W$. Then there exists open neighborhoods $\mathcal{V}_0$ and $\mathcal{V}_1$ of Q and a diffeomorphism*

$$\psi : \mathcal{V}_0 \longrightarrow \mathcal{V}_1$$

such that $\psi|_Q = \mathrm{Id}_Q$ and $\psi^\star \omega_1 = \omega_0$.

Remark II.3.2. It is not easy to create a diffeomorphism "ex nihilo". The remarkable idea of Moser is to construct *a whole path* of diffeomorphisms starting from the identity and ending at some diffeomorphism which has the desired property.

Let us write the proof of Moser lemma when $W = \mathbf{C}^n$ and explain then what should be done to get it in the general case (essentially to replace the Euclidean structure by a Riemannian metric). Consider the normal bundle to Q in $\mathbf{C}^n$,

$$NQ = \{(x, v) \in Q \times \mathbf{C}^n \mid v \perp T_x Q\}$$

and the open subset

$$\mathcal{U}_\varepsilon = \{(x, v) \in N_Q \mid \|v\| < \varepsilon\}.$$

Notice firstly that:

Lemma II.3.3. *Let Q be a compact submanifold of the Euclidean space $\mathbf{R}^m$. For ε small enough, the map*

$$
\begin{aligned}
E : NQ &\longrightarrow \mathbf{R}^m \\
(x, v) &\longmapsto x + v
\end{aligned}
$$

is a diffeomorphism from $\mathcal{U}_\varepsilon$ onto its image.

Proof. In a neighbourhood of a point x_0 of Q, we describe Q by local coordinates $u = (u_1, \ldots, u_k)$, namely by a mapping $x : U \to \mathbf{R}^m$ where U is open in $\mathbf{R}^k$ and $x(0) = x_0$. One can choose vector fields $(v_1(u), \ldots, v_{m-k}(u))$ of $\mathbf{R}^m$ on U, that form, for all u, an orthonormal basis of the normal space of Q at $x(u)$. So we have local coordinates $(u_1, \ldots, u_k, t_1, \ldots, t_{m-k})$ on NQ in which the mapping E is

$$E(u, t) = x(u) + \sum_{i=1}^{m-k} t_i v_i(u).$$

The partial derivatives are

$$
\begin{cases}
\dfrac{\partial E}{\partial u_i} = \dfrac{\partial x}{\partial u_i} + \sum_j t_j \dfrac{\partial v_j}{\partial u_i} \\[3mm]
\dfrac{\partial E}{\partial t_k} = v_k.
\end{cases}
$$

The matrix of partial derivatives is invertible for $t = 0$, thus it is invertible also for $\|t\|$ small enough[1]. We conclude globally using the compactness of Q. $\qquad\square$

Call $\mathcal{V}_0$ the image of a suitable $\mathcal{U}_\varepsilon$. This is a neighbourhood of Q in $\mathbf{C}^n$.

Lemma II.3.4. *On $\mathcal{V}_0$, the 2-form $\tau = \omega_1 - \omega_0$ is exact.*

---

[1]It is interesting to see "how far" we can go. This leads to the notion of *focal point*, see for example [26].

First proof. The vector bundle NQ retracts on its zero section. The inclusion $j : Q \to V_0$ thus induces an isomorphism $j^* : H^2_{DR}(V_0) \to H^2_{DR}(Q)$. As $j^*[\omega_1] = j^*[\omega_0]$, the cohomology classes of ω_1 and ω_0 are equal in $H^2_{DR}(V_0)$, which means that their difference is an exact form. □

Second proof. We explicitly construct a 1-form σ that is a primitive of τ. Consider the dilatation of factor t in the fibers

$$\varphi_t : \quad \begin{array}{ccc} V_0 & \longrightarrow & V_0 \\ x+v & \longmapsto & x+tv \end{array} \qquad t \in [0,1].$$

This is a diffeomorphism (onto its image) for $t > 0$ and we have $\varphi_0(V_0) = Q$, $\varphi_1 = \mathrm{Id}_{V_0}$ and $\varphi_t|_Q = \mathrm{Id}_Q$. The form $\tau = \omega_1 - \omega_0$ is a 2-form on $\mathbf{C}^n$. Consider its restriction to V_0. It is identically zero along Q by assumption. We have

$$\varphi_0^* \tau = 0, \quad \varphi_1^* \tau = \tau.$$

Consider now the (time depending) radial vector field X_t (tangent to the dilatation) on V_0. This is the vector field defined by

$$X_t(y) = \left(\frac{d}{ds}\varphi_s\right) \left(\varphi_t^{-1}(y)\right)|_{s=t}.$$

It is defined only for $t > 0$, in the same way that φ_t is a diffeomorphism only for $t > 0$. In a very concrete way, the vector field is

$$X_t(x+v) = \frac{1}{t}v.$$

For all t, consider also the 1-form σ^t defined by

$$\sigma^t_{x+v}(Y) = \tau_{x+tv}(v, T_{x+v}(\varphi_t)(Y)).$$

Notice that, if y is in Q, one has

$$\varphi_t(y) = y \text{ and } \frac{d}{dt}\varphi_t(y) = 0$$

thus σ^t is zero along Q. For $t > 0$, one has

$$
\begin{aligned}
(\varphi_t^* \iota_{X_t} \tau)_{x+v}(Y) &= (\iota_{X_t} \tau)_{x+tv}(X_t(x+tv), T_{x+v}(\varphi_t)(Y)) \\
&= \tau_{x+tv}(v, T_{x+v}(\varphi_t)(Y)) \\
&= \sigma^t_{x+v}(Y).
\end{aligned}
$$

Hence, for $t > 0$,

$$\sigma^t = \varphi_t^* \iota_{X_t} \tau$$

and consequently

$$
\begin{aligned}
d\sigma^t &= d\left(\varphi_t^\star \iota_{X_t} \tau\right) \\
&= \varphi_t^\star \left(d\iota_{X_t}\tau + \iota_{X_t} d\tau\right) \\
&= \varphi_t^\star \left(\mathcal{L}_{X_t}\tau\right) \\
&= \frac{d}{dt}\left(\varphi_t^\star \tau\right).
\end{aligned}
$$

Eventually, we get

$$
d\sigma^t = \frac{d}{dt}\left(\varphi_t^\star \tau\right)
$$

for $t > 0$ and thus also for all $t \in [0,1]$. Now

$$
\tau = \tau - 0 = \varphi_1^\star \tau - \varphi_0^\star \tau = \int_0^1 \frac{d}{dt}\left(\varphi_t^\star \tau\right) dt = \int_0^1 (d\sigma^t)dt = d\sigma
$$

writing $\sigma = \int_0^1 \sigma^t dt$. We has thus proved that, in a neighbourhood of Q, $\omega_1 - \omega_0 = d\sigma$ is an exact form (with σ identically zero on Q). □

To finish the proof of Lemma II.3.1, we use the actual method of Moser. We consider the path of symplectic forms

$$
\omega_t = \omega_0 + t(\omega_1 - \omega_0) = \omega_0 + td\sigma.
$$

For $t = 0$, this is the non degenerate form ω_0. Also, along Q, this is the very same form ω_0. Restricting again $\mathcal{V}_0$ if necessary (using compactness again) one can assume that ω_t is non degenerate on $\mathcal{V}_0$ for all $t \in [0,1]$. Let Y_t be the vector field defined by

$$
\iota_{Y_t} \omega_t = -\sigma
$$

(the existence and uniqueness of Y_t are consequences of the fact that ω_t is non degenerate). Let ψ_t be its flow:

$$
\frac{d}{dt}\psi_t = Y_t \circ \psi_t.
$$

We have

$$
\begin{aligned}
\frac{d}{dt}\left(\psi_t^\star \omega_t\right) &= \psi_t^\star \left(\frac{d}{dt}\omega_t + \mathcal{L}_{Y_t}\omega_t\right) \\
&= \psi_t^\star \left(d(\sigma) + d\iota_{Y_t}\omega_t\right) \\
&= d\left(\psi_t^\star(\sigma + \iota_{Y_t}\omega_t)\right) \\
&= 0
\end{aligned}
$$

by definition of Y_t. Hence $\psi_t^\star \omega_t = \psi_0^\star \omega_0 = \omega_0$ and eventually

$$
\psi_1^\star \omega_1 = \omega_0.
$$

□

Remark II.3.5. In a general symplectic manifold W, the proof is identical to the one given here; what we need is the notion of a normal bundle, that is, of orthogonality in TW, and a way to replace the mapping $(x, v) \mapsto x + v$. One uses a Riemannian metric on W and its exponential mapping: the point $\exp_v(x)$ that replaces $x + v$ is the point reached at time 1 by a geodesic[2] starting from x (at time 0) with tangent vector v.

The most direct application of Lemma II.3.1 is the Darboux theorem. This is the case where Q is a point x_0, ω_1 is a symplectic form on W and ω_0 is the symplectic form induced on $T_{x_0} W$.

Theorem II.3.6 (Darboux theorem). *Let x be a point of a manifold W endowed with a symplectic form ω. There exists local coordinates*

$$(x_1, \ldots, x_n, y_1, \ldots, y_n)$$

centered at x in which $\omega = \sum dy_i \wedge dx_i$.

Proof. The form induced by ω_1 on $T_{x_0} W$ defines, using a diffeomorphism from a neighbourhood of 0 in $T_{x_0} W$ onto a neighbourhood of x_0 in W, a symplectic form ω_0 on a neighbourhood of x_0. Lemma II.3.1 gives a diffeomorphism ψ from a neighbourhood of x_0 into itself, that fixes x_0 and satisfies $\psi^* \omega_1 = \omega_0$. By definition of ω_0, there exists local coordinates centered at x_0 in which it can be written $\sum dy_i \wedge dx_i$. $\qquad\square$

II.3.b Tubular neighborhoods

The next application is a theorem of Weinstein that describes the tubular neighborhoods of the Lagrangian submanifolds.

Theorem II.3.7 (Weinstein [34]). *Let (W, ω) be a symplectic manifold and let $L \subset W$ be a compact Lagrangian submanifold. There exists a neighbourhood $\mathcal{N}_0$ of the zero section in $T^* L$, a neighbourhood $\mathcal{V}_0$ of L in W and a diffeomorphism $\varphi : \mathcal{N}_0 \to \mathcal{V}_0$ such that*

$$\varphi^* \omega = -d\lambda \text{ and } \varphi|_L = \mathrm{Id}.$$

Proof. Let us check that we can apply Lemma II.3.1. The submanifold Q is the Lagrangian submanifold L and the form ω_0 is the restriction of ω. The form ω_1 is the symplectic form of $T^* L$. We are going to compare them in $T^* L$. As in the previous proof, let us assume firstly that $W = \mathbf{C}^n$. Let φ be the composed mapping

$$
\begin{array}{ccccc}
T^* L & \longrightarrow & N_L & \longrightarrow & \mathbf{C}^n \\
(x, \alpha) & \longmapsto & (x, J v_\alpha) & \longmapsto & x + J v_\alpha
\end{array}
$$

---

[2]To extend the geodesics, we also need an assumption on the completeness of the metric, or on the manifold W.

where

- $\alpha \mapsto v_\alpha$ is the isomorphism between cotangent and tangent spaces given by the Euclidean structure of $\mathbf{C}^n$ restricted to L:

$$\alpha(u) = (u, v_\alpha),$$

- J is the multiplication by i. Recall (see Lemma I.2.1) that L is Lagrangian if and only if $TL^\perp = JTL$.

Call $\mathcal{N}_0$ a neighbourhood of the zero section in T^*L, mapped onto a suitable $\mathcal{U}_\varepsilon$, so that $\varphi : \mathcal{N}_0 \to \mathbf{C}^n$ is a diffeomorphism onto its image. We want to compare, in $\mathcal{N}_0 \subset T^*L$, the two forms $\omega_1 = -d\lambda$ and $\omega_0 = \varphi^*\omega$. To apply Lemma II.3.1, we have to check that they coincide along the zero section.

Let $(x, 0) \in L \subset \mathcal{N}_0$. We have

$$T_{(x,0)}\mathcal{N}_0 = T_{(x,0)}(T^*L) = T_x L \oplus T_x^* L.$$

Recall that there is an exact sequence

$$0 \longrightarrow \operatorname{Ker} T_{(x,\alpha)}\pi \longrightarrow T_{(x,\alpha)}(T^*L) \xrightarrow{T_{(x,\alpha)}\pi} T_x L \longrightarrow 0$$

which splits along the zero section s, using

$$T_x L \xrightarrow{T_x s} T_{(x,0)}(T^*L),$$

and that the kernel $\operatorname{Ker} T_{(x,\alpha)}\pi$ is canonically identified with $T_x^* L$. Compute then $\varphi^*\omega$ along the zero section. For $v, w \in T_x L$ and $\alpha, \beta \in T_x^* L$, we have

$$\begin{aligned}
(\varphi^*\omega)_{(x,0)}((v,\alpha),(w,\beta)) &= \omega_{\varphi(x,0)}(v + Jv_\alpha, w + Jv_\beta) \\
&= \omega_x(v + Jv_\alpha, w + Jv_\beta) \\
&= (v, v_\beta) - (w, v_\alpha) \\
&= \beta(v) - \alpha(w).
\end{aligned}$$

But we have seen (in Exercise II.1) that

$$(d\lambda)_{(x,0)}((v,\alpha),(w,\beta)) = \left(\sum dy_j \wedge dx_j\right)((v,\alpha),(w,\beta)) = \alpha(w) - \beta(v).$$

The forms $\varphi^*\omega$ and $-d\lambda$ coincide along the zero section, therefore we can apply the lemma. $\square$

In the general situation where W is a symplectic manifold, we need a Riemannian metric and an analogue of J. We use an "almost complex structure" J calibrated by ω, namely an endomorphism J of the tangent bundle TW such that $J^2 = -\operatorname{Id}$ and

$$(X, Y) \longmapsto \omega(X, JY)$$

is a Riemannian metric. Such structures exist and form a contractible set. See for instance [5, 24]. Notice that this notion is a globalization of the linear notion, mentioned in §I.6.b.

II.3.c "Moduli space" of Lagrangian submanifolds

We consider now, for a given manifold L, the *space* of Lagrangian immersions

$$f : L \longrightarrow W.$$

We call it a "space" because this set is actually a topological space, a fact which allows to consider immersions that are "close" to a given immersion. We use the Whitney $\mathcal{C}^1$-topology.

The $\mathcal{C}^1$-topology

Let V and W be two manifolds. The $\mathcal{C}^1$-topology is a topology on the space of $\mathcal{C}^1$-maps from V to W. Consider the vector bundle $\mathcal{L}(TV, TW)$ over $V \times W$, the fiber at (x, y) of which is the vector space $\mathcal{L}(T_x V, T_y W)$. The total space is usually called $J^1(V, W)$ rather than $\mathcal{L}(TV, TW)$. Every map $f \in \mathcal{C}^1(V, W)$ defines a mapping

$$j^1 f : V \longrightarrow J^1(V, W)$$
$$x \longmapsto (x, f(x), T_x f).$$

If U is an open subset of $J^1(V, W)$, denote

$$\mathcal{V}(U) = \left\{ f \in \mathcal{C}^1(V, W) \mid j^1 f \in U \right\}.$$

The $\mathcal{C}^1$-topology is the topology for which the $\mathcal{V}(U)$ are a basis. It is said that a map f is "$\mathcal{C}^1$-close" to f_0 if it is close to f_0 for the $\mathcal{C}^1$-topology.

Diffeomorphism group

The group of diffeomorphisms of L acts on this space by $\varphi \cdot f = f \circ \varphi^{-1}$. We want to consider Lagrangian immersions only up to this action: we do not want to take into account the way the manifold L is "parametrized".

Moduli space

We consider the space of Lagrangian $\mathcal{C}^1$-immersions from L to W up to the action of the diffeomorphism group. The quotient space is called the "moduli space" of Lagrangian immersions from L to W and denoted $\mathcal{L}(L)$. The next theorem describes the Lagrangian immersions that are close to a fixed Lagrangian embedding of L into W.

Theorem II.3.8. *Let L be a compact and connected manifold. A neighbourhood of a Lagrangian embedding*

$$L \longrightarrow W$$

in the space $\mathcal{L}(L)$ can be identified with a neighbourhood of 0 in the vector space of closed 1-forms of class $\mathcal{C}^1$ on L.

Proof. Let $f_0 : L \to W$ be a Lagrangian embedding and $f : L \to W$ be a Lagrangian immersion close to f_0. In particular, f is close to f_0 for the "$\mathcal{C}^0$-topology[3]", we can consider that everything lies in a neighbourhood of L. Thanks to the tubular neighbourhood theorem (here Theorem II.3.7) we can assume that everything takes place in a neighbourhood of the zero section in T^*L. The map f is $\mathcal{C}^1$-close to the inclusion of the zero section $L \to T^*L$ (this is what f_0 has become when we have identified the neighbourhood of $f_0(L)$ in W with a neighbourhood of the zero section in T^*L). Thus the composition of f with the projection of the cotangent is a $\mathcal{C}^1$-mapping $L \to L$, close to the identity. Recall the next lemma, which is a consequence of the inverse function theorem.

Lemma II.3.9. *Let L be a compact and connected manifold. Let f be a $\mathcal{C}^1$-map $L \to L$ that is $\mathcal{C}^1$-close to the identity. Then f is a diffeomorphism.* □

According to this lemma, the composition is a diffeomorphism g of L. Composing with g^{-1}, we get an embedding

$$\alpha : L \longrightarrow T^*L$$

which is still $\mathcal{C}^1$-close to the zero section... but now the composition

$$L \longrightarrow T^*L \longrightarrow L$$

is the identity. Thus α is a section, that is, a 1-form on L, and α is closed because the embedding is Lagrangian. Conversely, all the closed 1-forms that are close to the zero section define Lagrangian embeddings close to f_0. □

Remark II.3.10. One should have noticed that the section $L \to T^*L$ defined by a 1-form is a $\mathcal{C}^1$-mapping if and only if the form is a $\mathcal{C}^1$-form. The $\mathcal{C}^1$-topology thus defines the structure of a topological vector space on the space of 1-forms. In §II.6 below, we will need a Banach space structure.

Remark II.3.11. The vector space we have obtained is infinite dimensional. It can be considered as a neighbourhood of f_0 in the "manifold" of deformations of f_0, or as its tangent space at f_0.

II.4   Calabi-Yau manifolds

We want now to describe, in a way analogous to what we have done in §II.3.b, the moduli space of special Lagrangian submanifolds. In order to apply Theorem II.3.7 (special Lagrangian submanifolds are, firstly, Lagrangian submanifolds) we need a *compactness* assumption on the Lagrangian submanifold. Unfortunately, as we have seen it in §I.5.b, the special Lagrangian submanifolds of $\mathbf{C}^n$ are never compact. We thus need to consider more general manifolds, in which it is possible to define special Lagrangian submanifolds. These are the "Calabi-Yau" manifolds.

---

[3]The $\mathcal{C}^0$-topology, defined similarly to the $\mathcal{C}^1$-topology, is simply the compact open topology.

The point is to define a structure that globalizes the structures on $\mathbf{C}^n$ which have allowed us to speak of special Lagrangian submanifolds. Recall that, in addition to the $\mathbf{R}$-bilinear alternated form ω, we have used the form $\Omega = dz_1 \wedge \cdots \wedge dz_n$ of the complex determinant.

We will use here the best adapted definition of a Calabi-Yau manifold, the point is not to spend time on the Calabi-Yau manifold itself but rather on its special Lagrangian submanifolds. For more information on Calabi-Yau manifolds, see [33, 8] and the references they contain.

II.4.a   Definition of the Calabi-Yau manifolds

Our manifolds should be complex and endowed with a symplectic form ω and a type-$(n, 0)$ holomorphic form Ω that is nowhere zero (this is sometimes called a holomorphic volume form). Consider thus a manifold M, on which are given

- a complex structure J (multiplication by i),

- a closed non degenerate type $(1, 1)$-form ω (the Kähler form)

- a Riemannian metric
$$g(X, Y) = \omega(X, iY),$$

- a Hermitian metric

$$h(X, Y) = g(X, Y) - i\omega(X, Y),$$

- a trivialization of the "canonical" bundle $\Lambda^n T^\star M$, namely a type-$(n, 0)$ holomorphic form Ω which is nowhere zero.

We still need a relation between the forms ω and Ω. Notice that both forms $\omega^{\wedge n}$ and $\Omega \wedge \bar{\Omega}$ are of type (n, n) and both do not vanish on M, in particular, both are volume forms. We thus have
$$\Omega \wedge \bar{\Omega} = f\omega^{\wedge n}$$

for some function f on M. The additional compatibility condition is that f should be constant. Let us look at the case of $\mathbf{C}^n$. We have

$$\omega^n = (\sum_{j=1}^{n} dy_j \wedge dx_j)^{\wedge n} = n!(dy_1 \wedge dx_1) \wedge \cdots \wedge (dy_n \wedge dx_n).$$

Writing
$$dy = \frac{1}{2i}(dz - d\bar{z}) \text{ and } dx = \frac{1}{2}(dz + d\bar{z})$$
and noticing that

$$dy \wedge dx = \frac{1}{4i}(dz - d\bar{z}) \wedge (dz + d\bar{z}) = \frac{1}{2i}dz \wedge d\bar{z},$$

we can also write

$$\omega^{\wedge n} = \frac{n!}{2^n i^n}\left(dz_1 \wedge d\bar{z}_1 \wedge \cdots \wedge dz_n \wedge d\bar{z}_n\right).$$

The computation of $\Omega \wedge \bar{\Omega}$ gives

$$\Omega \wedge \bar{\Omega} = (dz_1 \wedge \cdots \wedge dz_n) \wedge (d\bar{z}_1 \wedge \cdots \wedge d\bar{z}_n).$$

We thus have

$$\omega^{\wedge n} = \frac{(-1)^{\frac{n(n-1)}{2}} n!}{2^n i^n}\Omega \wedge \bar{\Omega}.$$

We will use the same normalization formula to define a Calabi-Yau manifold in general.

Definition II.4.1. A complex manifold M is said to be a *Calabi-Yau manifold* if it is Kähler, has a trivialized canonical bundle, and if the Kähler form ω and the type-$(n,0)$ form Ω trivializing the bundle $\Lambda^n T^\star M$ are related by

$$\omega^{\wedge n} = \frac{(-1)^{\frac{n(n-1)}{2}} n!}{2^n i^n}\Omega \wedge \bar{\Omega}.$$

Remark II.4.2. Recall that it is possible to express the fact that the form ω is Kähler by saying that the complex structure is "parallel" with respect to the Levi-Cività connection associated with the metric it defines with ω. Similarly, it is possible to express the compatibility condition for Ω by saying that it is parallel with respect to the same connection.

Remark II.4.3. In general, it is required that the Kähler metric be *complete*, in other words that it is possible to extend geodesics. This is equivalent to requiring that the manifold be complete (in the sense of metric spaces).

II.4.b Yau's theorem

Consider a (complex algebraic) projective smooth manifold M of complex dimension n. Assume that all the $H^{p,0}(M)$ are zero for $1 \leq p \leq n-1$ and that the canonical bundle $\Lambda^n T^\star M = K_M$ is trivialized by a type-$(n,0)$ form Ω. Notice that M is Kähler, call the Kähler form ω. Rescaling ω if necessary, we get

$$\int_M \omega^{\wedge n} = \frac{(-1)^{\frac{n(n-1)}{2}} n!}{2^n i^n}\int_M \Omega \wedge \bar{\Omega}.$$

A hard theorem of Yau [35] asserts that there exists a unique Kähler form $\widetilde{\omega}$ on M such that $[\widetilde{\omega}] = [\omega] \in H^2_{DR}(M)$ and which, together with Ω, gives M the structure of a Calabi-Yau manifold.

II.4.c Examples of Calabi-Yau manifolds

Of course $\mathbf{C}^n$ is a Calabi-Yau manifold.

Affine quadrics

We have defined in §II.1 a symplectic form on the unit sphere $S^2 \subset \mathbf{R}^3$ by the formula

$$\omega_x(X, X') = \det(x, X, X').$$

Similarly, the formula

$$\Omega_z(Z, Z') = \det_{\mathbf{C}}(z, Z, Z')$$

defines a "holomorphic symplectic" form of the complex quadric

$$Q = \left\{ (z_1, z_2, z_3) \mid z_1^2 + z_2^2 + z_3^2 = 1 \right\}.$$

In "differential" terms,

$$\Omega = z_1 dz_2 \wedge dz_3 + z_2 dz_3 \wedge dz_1 + z_3 dz_1 \wedge dz_2.$$

On the open subset of Q where $z_3 \neq 0$, z_1 and z_2 are coordinates and, using the relation

$$z_1 dz_1 + z_2 dz_2 + z_3 dz_3 = 0,$$

we can write

$$\Omega = \frac{1}{z_3} dz_1 \wedge dz_2,$$

so that

$$\Omega \wedge \bar{\Omega} = \frac{1}{|z_3|^2} dz_1 \wedge dz_2 \wedge d\bar{z}_1 \wedge d\bar{z}_2.$$

Modifying the restriction ω_0 to Q of the standard Kähler form of $\mathbf{C}^3$, let us construct a Kähler form ω on Q such that

$$\omega \wedge \omega = \frac{1}{4} \Omega \wedge \bar{\Omega}.$$

Call h the restriction to Q of the function $|z|^2$. We look for ω of the form

$$\omega = \frac{i}{2} \partial \bar{\partial} (f \circ h)$$

for some function f. A straightforward computation (see also [31]) shows that $f(h) = \sqrt{h+1}$ works.

The quadric Q, equipped with Ω and ω is (thus) a Calabi-Yau manifold. Recall that Q is diffeomorphic to the tangent bundle TS^2 by

$$Q \longrightarrow TS^2$$

$$X + iY \longmapsto \left(\frac{X}{\sqrt{1 + \|Y\|^2}}, Y \right).$$

In this way, what we have got is the structure of a Calabi-Yau manifold on the tangent (or cotangent) bundle of the sphere S^2. It is possible (but a little more complicated) to do the same for the cotangent bundles of all the spheres S^n and more generally for those of all "rank-1 symmetric spaces" (see [31]).

Remark II.4.4. Recall that we have identified $\mathbf{C}^2$ with the skew field $\mathbf{H}$ of quaternions (in §I.5.a). Similarly, the surface Q has the structure of a "quaternionic" or "hyperkähler" manifold.

Call I the complex structure defined on Q by that of $\mathbf{C}^3$ (this is the multiplication by i) and notice that the symmetric bilinear form that is an equation for Q is still non degenerate when restricted to $z^\perp = T_z Q$. Define an operator J_z on the tangent space $T_z Q$ by the fact that $J_z(Z)$ is the unique vector in $T_z Q$ that is orthogonal to Z for the complex bilinear form and such that

$$\det{}_\mathbf{C}(z, Z, J_z(Z)) = \|Z\|^2.$$

This is an almost complex structure since

$$\det{}_\mathbf{C}(z, J_z Z, -Z) = \|Z\|^2$$

thus $J_z^2 = -\,\mathrm{Id}$. This is an isometry since

$$\|JZ\|^2 = \det{}_\mathbf{C}(z, JZ, J^2(Z)) = \det{}_\mathbf{C}(z, JZ, -Z) = \|Z\|^2.$$

Moreover, J "anti-commutes" with I:

$$\det{}_\mathbf{C}(z, IZ, JIZ) = \|IZ\|^2 = \|Z\|^2 \text{ on the one hand}$$
$$= i \det{}_\mathbf{C}(z, Z, JIZ) \text{ by linearity.}$$

We thus have

$$\det{}_\mathbf{C}(z, Z, JIZ) = -i \|Z\|^2 = -\det{}_\mathbf{C}(z, Z, IJZ)$$

so that $JI = -IJ$. Hence I, J and IJ form a quaternionic structure on Q. On Q, we thus have

- the Kähler form ω,

- the complex structure I defined by multiplication by i in $\mathbf{C}^3$,

- the associated Riemannian metric g, so that $\omega(X, IY) = g(X, Y)$,

- the "holomorphic symplectic form" Ω,

- the complex structure J defined in such a way that Ω be a J-Kähler form, associated with the same metric g.

It is said that Q is hyperkähler. See Exercise II.7 for a kind of converse statement.

Let us give now a few examples of compact Calabi-Yau manifolds.

Elliptic curves

The quotient M of $\mathbf{C}$ by a lattice Λ is an *elliptic curve*. The two forms

$$\omega = \frac{1}{2i} dz \wedge d\bar{z} \text{ and } \Omega = dz$$

give it the structure of a dimension-1 Calabi-Yau manifold. One can, more generally, perform the quotient of $\mathbf{C}^n$ by a lattice. It is time for a remark: no other "explicit" example of compact Calabi-Yau manifold is known. In all the known examples, the existence of the Kähler metric with all the desired properties is obtained as a consequence of the Yau theorem (§ II.4.b).

Hypersurfaces

Recall that complex elliptic curves can be considered as degree-3 curves in $\mathbf{P}^2(\mathbf{C})$, thanks to the Weierstrass $\wp$-function. They are thus the $n = 1$ case in the next theorem.

Theorem II.4.5. *A degree-d hypersurface in $\mathbf{P}^{n+1}(\mathbf{C})$ is a dimension-n Calabi-Yau manifold if and only if $d = n + 2$.*

Proof. The condition on the degree is necessary, as we show it now by the computation of the first Chern classes. We want that the bundle $\Lambda^n T^\star M$ be trivializable, we must thus have $c_1(T^\star M) = -c_1(TM) = 0$. Calling j the inclusion of M in $\mathbf{P}^{n+1}(\mathbf{C})$, we have

$$c_1(TM) + j^\star c_1(\mathcal{O}(d)) = j^\star c_1(TP^{n+1}(\mathbf{C}))$$

since the normal bundle of M in $\mathbf{P}^{n+1}(\mathbf{C})$ is $\mathcal{O}(d)$. Denoting by t the dual class to the hyperplane section in $H^2(\mathbf{P}^{n+1}(\mathbf{C}))$, we have

$$(n + 2 - d)j^\star t = 0$$

so that $d = n + 2$.

Assume conversely that $d = n + 2$. Let us construct explicitly a holomorphic n-form on M. Let F be a degree-$(n + 2)$ homogeneous polynomial that describes the hypersurface M. Every point of M lies in an affine chart $Z_i \neq 0$ of $\mathbf{P}^{n+1}(\mathbf{C})$. In affine coordinates $z_k = Z_k/Z_i$, there is an index j such that

$$\frac{\partial}{\partial z_j} F(z_0, \ldots, 1, \ldots, z_{n+1}) \neq 0$$

since M is smooth. The formula

$$\Omega = (-1)^{i+j-1} \frac{dz_0 \wedge \cdots \wedge \widehat{dz_i} \wedge \cdots \wedge \widehat{dz_j} \wedge \cdots \wedge dz_{n+1}}{\frac{\partial F}{\partial z_j}(z_0, \ldots, 1, \ldots, z_{n+1})}$$

defines a homogeneous holomorphic n-form on M that is nowhere zero. This is a consequence of the theorem of Yau (§ II.4.b) that there is, indeed, in the same cohomology class as the standard Kähler form ω, another Kähler form $\omega + i\partial\bar{\partial}\varphi$ giving a Calabi-Yau structure on M. $\qquad\qquad\square$

Remark II.4.6. The form Ω above is defined as "Poincaré residue[4]" starting from the $n+1$-form on $\mathbf{P}^{n+1}(\mathbf{C})$ with poles along M defined by

$$\sigma_i = (-1)^i \frac{dz_0 \wedge \cdots \wedge \widehat{dz_i} \wedge \cdots \wedge dz_{n+1}}{F(z_0, \ldots, 1, \ldots, z_{n+1})}$$

in the affine chart $Z_i \neq 0$.

Remark II.4.7. Calabi-Yau manifolds of dimension 2 are hyperkähler. The proof of this fact is the subject of Exercises II.6 and II.7.

II.4.d Special Lagrangian submanifolds

An immersion $f : V \to M$ from a manifold of real dimension n into a Calabi-Yau manifold M of complex dimension n is said *special Lagrangian* if it satisfies $f^\star\omega = 0$ and $f^\star\beta = 0$. As in the case of $\mathbf{C}^n$, the form $f^\star\Omega = f^\star\alpha$ is then a volume form.

II.5 Special Lagrangians in real Calabi-Yau manifolds

II.5.a Real manifolds

A complex analytic manifold is *real* if it is endowed with a "real structure", that is, with an anti-holomorphic involution S : an involution such that, for any holomorphic function f over an open subset U of M, $\overline{f \circ S}$ is a holomorphic function. For example, on the algebraic submanifolds of $\mathbf{P}^N(\mathbf{C})$ described by real polynomial equations, the complex conjugation is an anti-holomorphic involution. These manifolds are thus real manifolds. In particular, the projective space $\mathbf{P}^N(\mathbf{C})$ itself is a real manifold.

The real part, or set of real points of a real manifold is, by definition, the set of fixed points of S. For example, the real part of the real manifold $\mathbf{P}^N(\mathbf{C})$ is $\mathbf{P}^N(\mathbf{R})$. Notice that there exists respectable real manifolds that have no real point at all, as is, for example, the "Euclidean quadric"

$$\sum_{i=1}^{N+1} X_i^2 = 0$$

in $\mathbf{P}^N(\mathbf{C})$.

---

[4]See [15] p. 147.

Proposition II.5.1. *The real part of a real manifold of complex dimension n, if it is non empty, is a submanifold all connected components of which have dimension n.*

Proof. The connected components of the set of fixed points of the action of a finite group (here the order-2 group generated by S) are always submanifolds. The tangent space at x to such a component is the subspace of fixed points of the **R**-linear involution $\sigma = T_x S$.

The fact that S is a real structure implies that $\overline{f \circ \sigma}$ is a complex linear form for any complex linear form f on the tangent space at x. We have to check that the fixed subspace of σ has dimension n. To do this, we simply verify that the eigensubspaces associated with the eigenvalues 1 and -1 are isomorphic. Indeed, if $\sigma(X) = X$, then for any *complex* linear form f, we have

$$\overline{f \circ \sigma}(iX) = \overline{i \overline{f} \circ \sigma}(X) = i\overline{f}(X) = \overline{f}(-iX).$$

For any complex linear form f, we thus have

$$f(\sigma(iX)) = f(-iX)$$

so that $\sigma(iX) = -iX$. Hence, there are "as many" eigenvectors for the eigenvalue -1 than there are for the eigenvalue 1. $\qquad\square$

II.5.b Real Calabi-Yau manifolds

A Calabi-Yau manifold is *real* if it is both a Calabi-Yau manifold and a real manifold, with a couple of compatibility conditions

$$S^\star \omega = -\omega \text{ and } S^\star \Omega = \bar{\Omega}$$

(similarly to what happens in $\mathbf{C}^n$ with the complex conjugation and the two usual forms Ω and ω).

Examples II.5.2. • The affine quadric $\sum z_i^2 = 1$ of $\mathbf{C}^3$, endowed with the complex conjugation of coordinates is a real manifold. It is also clear that this is a real Calabi-Yau manifold. Its real part is simply the unit sphere $S^2 \subset \mathbf{R}^3$. If we consider Q as the tangent bundle to S^2, notice that the complex conjugation is the multiplication by -1 on the fibers and that the real part is the zero section.

• A real hypersurface of degree $n+2$ in $\mathbf{P}^{n+1}(\mathbf{C})$ is a real Calabi-Yau manifold. This is checked by computing $S^\star \Omega$ and $S^\star(\omega + i\partial\bar{\partial}\varphi)$, for S the involution induced by the real structure (complex conjugation) of $\mathbf{P}^{n+1}(\mathbf{C})$ and Ω, ω as in the proof of Theorem II.4.5.

II.5.c   The example of elliptic curves

Let us come back to the example of $\Gamma = \mathbf{C}/\Lambda$ where Λ is a lattice that we assume here to have the form

$$\Lambda = \{m + n\tau \mid m, n \in \mathbf{Z}\}$$

for some fixed τ such that $0 \leq \operatorname{Re}(\tau) < 1$ et $\operatorname{Im}(\tau) > 0$. To define a real structure on $\mathbf{C}/\Lambda$ from the complex conjugation in $\mathbf{C}$, it is necessary that Λ be invariant, that is, that

$$\bar{\tau} = m + n\tau$$

for some $m, n \in \mathbf{Z}$. Considering the real and imaginary parts of τ, it is seen that $m = 2\operatorname{Re}(\tau)$, thus $\operatorname{Re}(\tau) = \frac{1}{2}$ or 0.

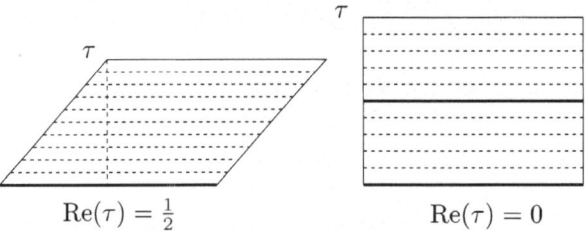

$$\operatorname{Re}(\tau) = \tfrac{1}{2} \qquad\qquad\qquad \operatorname{Re}(\tau) = 0$$

Figure II.1: Real elliptic curves

In the second case, the real part of Γ has two connected components, but in the first case, it has only one, as can be seen solving the equation

$$\bar{z} = z + m + n\tau$$

in both cases. These components are depicted in bold on Figure II.1.

Notice (although this is a trivial remark) that the lines that are parallel to the x axis constitute a real foliation of $\mathbf{C}/\Lambda$ by circles (dimension-1 tori) that are special Lagrangian submanifolds of Γ, represented by dotted lines on Figure II.1. The space of these special Lagrangian submanifolds is parametrized by the axis generated by τ or rather by its image in Γ, a circle.

We shall see more generally in §II.6 that the moduli space of special Lagrangian submanifolds in a Calabi-Yau manifold is, in the neighbourhood of a submanifold V, a manifold whose dimension is the first Betti number of V (here V is a circle and its first Betti number is 1).

II.5.d   Special Lagrangians in real Calabi-Yau manifolds

Assume now that M is a *real* Calabi-Yau manifold. We know that

$$S^{\star}\omega = -\omega \text{ and } S^{\star}\Omega = \bar{\Omega}.$$

Assume now that the real part $M_{\mathbf{R}}$ is not empty. Call j the inclusion of $M_{\mathbf{R}}$ into M. We have $S \circ j = j$ and in particular

$$j^* \omega = (S \circ j)^* \omega = j^* (S^* \omega) = j^* (-\omega) = -j^* \omega$$

hence $j^* \omega = 0$. Similarly

$$j^* \Omega = (S \circ j)^* \Omega = j^* (S^* \Omega) = j^* \bar{\Omega} = \overline{j^* \Omega}$$

thus $j^* \beta = 0$. We have proved:

Proposition II.5.3. *Let M be a real Calabi-Yau manifold. The real part of M, if it is not empty, is a special Lagrangian submanifold of M.* $\square$

Let us describe now a few examples of this situation.

The affine quadric

The sphere S^2 is a special Lagrangian submanifold of the affine quadric $Q \in \mathbf{C}^3$. In other words, with the Calabi-Yau structure on TS^2 defined in §II.4.c , the zero section is a special Lagrangian submanifold.

In the next examples, we consider a smooth hypersurface defined by a real homogeneous polynomial of degree $n + 2$ in $\mathbf{P}^{n+1}(\mathbf{C})$, with its real Calabi-Yau structure.

Elliptic curves

The $n = 1$ case, that of plane cubics, is isomorphic to the example of quotients of $\mathbf{C}$ by lattices (§ II.5.c). The real part of a plane cubic has zero, one or two connected components (see Figure II.1). All components are (topologically) circles. Cubics are foliated by special Lagrangian circles, drawn in dotted lines on Figure II.1.

Degree-4 surfaces

Consider now real algebraic surfaces of degree 4 in $\mathbf{P}^3(\mathbf{C})$ (the real part of this subject has been investigated and explained in [21]). Here is an example from [9]. Consider the real polynomial

$$P(z_0, z_1, z_2, z_3) = z_0^4 + z_1^4 - z_2^4 - z_3^4$$

that describes a smooth surface M which has a non empty real part $M_{\mathbf{R}}$. This real part is

$$M_{\mathbf{R}} = \left\{ (x_0, x_1, x_2, x_3) \in \mathbf{R}^4 - \{0\} \mid x_0^4 + x_1^4 = x_2^4 + x_3^4 \right\} / (x \sim \lambda x).$$

Normalize the non zero vectors of $\mathbf{R}^4$ by the choice, in each real line through zero, of one of the two vectors such that

$$x_0^4 + x_1^4 + x_2^4 + x_3^4 = 2.$$

Then $M_{\mathbf{R}}$ is the quotient

$$\left\{(x_0, x_1), (x_2, x_3) \in \mathbf{R}^2 \times \mathbf{R}^2 \mid x_0^4 + x_1^4 = x_2^4 + x_3^4 = 1\right\} / (u, v) \sim (-u, -v).$$

It is clear that the curve C described in $\mathbf{R}^2$ by the equation $x^4 + y^4 = 1$ is diffeomorphic to a circle (radially). Eventually

$$M_{\mathbf{R}} = (C \times C)/((u, v) \sim (-u - v))$$

is diffeomorphic to a torus. We have thus found a special Lagrangian torus in the Calabi-Yau surface M.

II.6   Moduli space of special Lagrangian submanifolds

We want now an analogue of Theorem II.3.8, more precisely a description of a neighbourhood of a given special Lagrangian submanifold in the space of all special Lagrangian submanifolds.

Theorem II.6.1 (McLean [25]). *Let V be a compact manifold. The moduli space of the special Lagrangian embeddings of V in the Calabi-Yau manifold M is a manifold of finite dimension $b_1(V) = \dim H^1(V; \mathbf{R})$. Its tangent space at a given point is isomorphic to the vector space of harmonic 1-forms on V.*

As usual, it is understood that the empty set is a manifold of any dimension — this is not an existence theorem.

Remark II.6.2. There are two main differences between this statement and Theorem II.3.8. The first one is that the moduli space here has finite dimension. The second one is that the condition "to be special Lagrangian" is no longer linear, so that this is indeed the tangent space that is identified to a space of differential forms.

Example II.6.3. Let us come back to the example of the Calabi-Yau structure on TS^2 described in § II.4.c. We have said in § II.5.d that the zero section is a special Lagrangian submanifold. As there are no non zero harmonic 1-forms on S^2, the theorem of McLean asserts that the zero section is "rigid", that is, it cannot be deformed. In the moduli space of special Lagrangian submanifolds, this is an isolated point.

Proof of Theorem II.6.1.
Using the tubular neighbourhood theorem (here in § II.3.b), replace M by a tubular neighbourhood of the submanifold V that is isomorphic to a neighbourhood of the zero section in the normal bundle of V (as are all tubular neighborhoods) and to a neighbourhood of the zero section in the cotangent T^*V. We will use the structures induced by those of M on this neighbourhood, keeping their names, for example $\Omega = \alpha + i\beta$.

We have said in §II.3.c that the space of Lagrangian submanifolds can be identified with a neighbourhood of 0 in the space $Z^1(V)$ of the closed 1-forms on V. The special Lagrangian submanifolds are described, in this space, by the equation $F(\eta) = 0$ where

$$F : Z^1(V) \longrightarrow \Omega^n(V)$$

is the mapping defined by $F(\eta) = \eta^\star \beta$. Although these spaces are infinite dimensional, the strategy of the proof is to show that F is submersive at 0. It is thus better to restrict, as much as possible, its target space. Notice first:

Lemma II.6.4. *The image of F is contained in the subspace $d\Omega^{n-1}(V)$ of exact n-forms on V.*

Proof of Lemma II.6.4. If η is the zero form, the mapping $\eta : V \to T^\star V$ is the inclusion of the zero section, a special Lagrangian, thus $F(0) = 0$. Given a form η, it is possible to consider the path (segment) $(t\eta)_{t \in [0,1]}$ joining it to the zero form... and giving a homotopy from the section η to the zero section. The cohomology class of the closed form $(t\eta)^\star \beta$ does not depend on t, thus it is identically zero, and that means, indeed, that $\eta^\star \beta$ is an exact form. □

We thus consider F as a mapping

$$F : Z^1(V) \longrightarrow d\Omega^{n-1}(V)$$

and compute its differential at 0.

Lemma II.6.5. *The differential of F at 0 is the mapping*

$$(dF)_0(\eta) = d(\star \eta)$$

where $\star$ denotes the Hodge star operator[5] associated with the metric defined by the Calabi-Yau structure on the special Lagrangian V.

Proof of Lemma II.6.5. To compute $(dF)_0(\eta)$, one chooses a path of forms $\tilde{\eta}_t$ whose tangent vector at 0 is the form η. Let η be a 1-form on V and X be the vector field that corresponds to it via the metric on V, that is, the vector field such that $g(X, \cdot) = \eta$. Let $Y = JX$ be the vector field normal to V. This is the vector corresponding to η under the isomorphism $NV \simeq T^\star V$. The vector field Y is only defined along V, we extend it (arbitrarily) in a vector field $\tilde{Y}$ on the tubular neighbourhood under consideration. Call $\tilde{\varphi}_t$ the flow of $\tilde{Y}$, so that $\tilde{\varphi}_t$ is a diffeomorphism defined for t small enough. The restriction φ_t of $\tilde{\varphi}_t$ to V is an embedding of V into NV (one pushes V using φ_t). For $t = 0$, this is the zero section. Hence for t small enough, this is still a section of NV. We have, for all x in V,

$$\frac{d}{dt}\varphi_t(x)|_{t=0} = Y(x).$$

---

[5]See Exercise II.5.

Under the identification $NV \simeq T^\star V$, the section φ_t of NV corresponds to a section $\widetilde{\eta}_t$ of $T^\star V$ which is a path of forms, whose tangent vector at 0 is the form η.

Consider now the $(n, 0)$-form Ω, still on our neighbourhood of the zero section in NV. We have

$$\frac{d}{dt}\widetilde{\varphi}_t^\star \Omega|_{t=0} = \mathcal{L}_{\widetilde{Y}}\Omega = d\iota_{\widetilde{Y}}\Omega$$

applying Cartan formula together with the fact that Ω is closed. For the embedding $\varphi_t : V \to NV$, we thus have

$$\frac{d}{dt}\varphi_t^\star \Omega|_{t=0} = d\iota_{JX}\Omega = id(\iota_X\Omega)$$

since Ω is **C**-linear. We then have

$$\begin{aligned}
(dF)_0(\eta) &= \frac{d}{dt}\widetilde{\eta}_t^\star \beta|_{t=0} \\
&= \mathrm{Im}(id(\iota_X\Omega)) \\
&= \mathrm{Re}\, d(\iota_X\Omega) \\
&= d(\iota_X\alpha).
\end{aligned}$$

We still have to convince ourselves that $\iota_X\alpha = \star\eta$. The $(n-1)$-form $\star\eta$ is the unique form satisfying

$$\psi \wedge (\star\eta) = g(\psi, \eta)\alpha$$

for any 1-form ψ. But, as the $(n+1)$-form $\psi \wedge \alpha$ is zero, its interior product by X is also zero and we have

$$\psi \wedge (\iota_X\alpha) = (\iota_X\psi)\alpha = g(\psi, \eta)\alpha$$

by definition of X and of the metric g on the space of 1-forms. $\square$

Notice that this implies in particular that the differential dF_0 is onto: if σ is an $(n-1)$-form,

$$d\sigma = (dF_0)(\pm \star \sigma).$$

To end the proof of the theorem, we need to precise what kind of implicit function theorem we use to go from "differential is surjective" to "inverse image is a submanifold". The simplest here is to use the standard implicit function theorem for Banach spaces (see, for example, [10]). We need to endow the spaces of forms $Z^1(V)$ and $\Omega^n(V)$ with structures of Banach spaces. Let us precise the regularity of the forms we use. We consider forms of class $\mathcal{C}^{1,\varepsilon}$ in $Z^1(V)$ and of class $\mathcal{C}^{0,\varepsilon}$ in $\Omega^n(V)$. The Hölder norm used here on forms is deduced from the usual Hölder norm on functions: recall that $\mathcal{C}^{k,\varepsilon}(U)$ is the space of functions of class $\mathcal{C}^k$ on the open set U of $\mathbf{R}^n$ all the derivatives (of order $\leq k$) of which have a finite Hölder norm $\|u\|_\varepsilon$ (for $\varepsilon \in]0, 1]$), with

$$\|u\|_\varepsilon = \sup_{x,y \in U} \frac{|u(x) - u(y)|}{\|x - y\|^\varepsilon} + \sup_{x \in U} |u(x)|.$$

The implicit function theorem gives the fact that $F^{-1}(0)$ is a submanifold in a neighbourhood of 0, whose tangent space at 0 is the kernel $\mathcal{H}^1(V)$ of $(dF)_0$. It is important here that this kernel has finite dimension. The isomorphism between $H^1_{DR}(V)$ and the space $\mathcal{H}^1(V)$ of harmonic 1-forms is the contents in degree 1 of the Hodge theorem, see [15]. $\qquad\qquad\qquad\qquad\qquad\qquad\qquad\qquad\qquad\qquad\qquad\square$

Remark II.6.6. The vector space $H^{n-1}_{DR}(V)$ is isomorphic to the vector space dual to $H^1_{DR}(V)$, so that

$$H^1_{DR}(V) \oplus H^{n-1}_{DR}(V)$$

has a natural symplectic structure (see Exercise I.1), here

$$\omega((\alpha, \eta), (\alpha', \eta')) = \int_V (\alpha \wedge \eta' - \alpha' \wedge \eta).$$

The space of harmonic 1-forms is a Lagrangian subspace, by

$$\mathcal{H}^1(V) \longrightarrow H^1_{DR}(V) \oplus H^{n-1}_{DR}(V)$$
$$\alpha \longmapsto (\alpha \quad , \quad \star\alpha)$$

(this is the graph of the mapping $\star$, which is symmetric with respect to the metric... see Exercise I.8). If $j_0 : V \to W$ is a special Lagrangian submanifold, call $\mathcal{B}$ the moduli space in a neighbourhood of j_0. We thus have a Lagrangian subspace

$$T_{j_0}\mathcal{B} \longrightarrow H^1_{DR}(V) \oplus H^{n-1}_{DR}(V)$$

and it is possible to "integrate" it in a Lagrangian embedding (see [19])

$$F : \mathcal{B} \longrightarrow H^1_{DR}(V) \oplus H^{n-1}_{DR}(V).$$

See also [11] for a description of all these structures by symplectic reduction.

II.7 Towards mirror symmetry?

The "mirror conjecture" asserts the existence, for any Calabi-Yau manifold M, of another Calabi-Yau manifold $M^\star$ of the same dimension, related with M in a way we briefly describe now, sending the readers to [33] for missing detail.

Call $\mathcal{M}_M$ the space of isomorphism classes of

- a complex structure J_t deforming the complex structure J of M

- a "complexified Kähler class" on (M, J_t), namely a cohomology class of the form $\alpha + i\beta$, for some Kähler class (for J_t) $\alpha \in H^2_{DR}(M)$ and some element $\beta \in H^2_{DR}(M)/2\pi H^2(M; \mathbf{Z})$.

Notice that, locally, $\alpha + i\beta$ varies in an open subset of $H^2(M; \mathbf{C})$, so that the space $\mathcal{M}_M$ is, locally, a product. The manifold M and its "mirror" partner $M^\star$ should be related by an isomorphism of the moduli spaces

$$\mathcal{M}_M \longrightarrow \mathcal{M}_{M^\star}$$

that exchanges the factors of this local decomposition as a product.

Using in an essential way the symplectic structure of the loop space of M and techniques that go far beyond the level of these notes, Givental has proved the conjecture in [14], following a series of previous papers, the references of which can be found in [14] and [33].

Special Lagrangian submanifolds have been a few years ago the central object of another approach to mirror symmetry, more speculative and having given so far very few results — but a very beautiful approach indeed, that I intend to describe very briefly here.

II.7.a   Fibrations in special Lagrangian submanifolds

We are no more interested in a single special Lagrangian submanifold but in a whole family. More precisely, we consider a compact Calabi-Yau manifold M and a differential mapping

$$p : M \longrightarrow B$$

to a manifold B, whose general fibers are special Lagrangian submanifolds. The dimension of B, as that of the fibers of p, must be n. It is not required that p be everywhere regular. Some of the fibers may be singular. The other ones, who correspond to regular values of p, are called *general* fibers.

We know (see §I.6.c and [4]) that in any proper Lagrangian fibration, the general fibers are unions of tori, so this must be the case here. The first Betti number of a torus of dimension n is precisely n, so that it can be expected that B "looks like" the moduli space of special Lagrangian submanifolds.

So, let $b \in B$ be a regular value of p and let $V \subset p^{-1}(b)$ be a connected component of the fiber $p^{-1}(b)$. If $X \in T_b B$ is a tangent vector, there exists a unique vector field Y normal to V in M and such that, for all x in V,

$$T_x p(Y_x) = X.$$

To this field Y corresponds a harmonic 1-form η on V, as in the proof of the theorem of McLean (here Theorem II.6.1). As B has dimension n, starting from n independent vectors $X_1, \ldots, X_n$ in $T_b B$, one constructs n fields $Y_1, \ldots, Y_n$, that are normal to V and linearly independent at each point of V. Dually, we thus have n harmonic 1-forms $\eta_1, \ldots, \eta_n$ that form a basis of $\mathcal{H}^1(V)$ and are linearly independent at each point of V.

In order that such a fibration $p : M \to B$ exists in a neighbourhood of a special Lagrangian torus $V \subset M$, it is necessary that, for the metric induced by the

Calabi-Yau structure on V, there exists a basis of $\mathcal{H}^1(V)$ consisting of forms that are independent at each point of V.

It is time to mention that (except in dimension 1) there is no known example having all the properties mentioned here.

- Notice first that, abstractly, a basis of harmonic 1-forms that are independent at each point exists on the flat torus, the basis $dx_1, \ldots, dx_n$ having this property. The metrics that are close enough to the flat metric thus have the same property.

- We have seen in §II.5.c that the situation of a Calabi-Yau manifold foliated by special Lagrangians submanifolds occurs in dimension 1.

- In dimension 2, on a special Lagrangian torus, one always has a basis of harmonic 1-forms as expected. We have seen that a special Lagrangian submanifold in dimension 2 is simply a complex curve (for a different complex structure). Assuming the submanifold is a torus, it must be an elliptic curve and it has a nowhere vanishing holomorphic form. Actually, the real and imaginary part of this form are harmonic forms on V and they are independent at every point.

II.7.b Mirror symmetry

The Strominger, Yau and Zaslow approach to mirror symmetry [32] is to associate, to a Calabi-Yau manifold M endowed with a fibration in special Lagrangian tori (assuming it exists), another Calabi-Yau manifold $M^\star$. The latter should be the "extended" moduli space of special Lagrangian submanifolds of M equipped with a flat unitary line bundle. Call, as above, $\mathcal{B}$ the moduli space of special Lagrangian submanifolds in the neighbourhood of V. Locally, the extended moduli space is

$$M^\star = \mathcal{B} \times H^1(V; \mathbf{R}/\mathbf{Z}).$$

Its tangent space at a point m is

$$T_m M^\star = H^1(V; \mathbf{R}) \oplus H^1(V; \mathbf{R}) \simeq H^1(V; \mathbf{R}) \otimes \mathbf{C}.$$

Thus, $M^\star$ has a natural almost complex structure, it is even Kähler:

Theorem II.7.1 (Hitchin [19]). *The complex structure on $M^\star$ is integrable, the metric of $H^1(V; \mathbf{R})$ defines a Kähler metric on $M^\star$.*

We have seen (Remark II.6.6 above) that $\mathcal{B}$ is a Lagrangian submanifold of $H^1_{DR}(V) \oplus H^{n-1}_{DR}(V)$, a symplectic vector space endowed by the metric of an almost complex structure (see Exercise I.12). It can be shown (see [19]) that $M^\star$ is a Calabi-Yau manifold if $\mathcal{B}$ is... a special Lagrangian submanifold in this complex vector space. See [11, 20].

Exercises

Exercise II.1. Check that the Liouville form λ of the cotangent T^*V satisfies

$$(d\lambda)_{(x,0)}((v,\alpha),(w,\beta)) = \alpha(w) - \beta(v)$$

(see Exercise I.1).

Exercise II.2. Let $\varphi : L \to L$ be a diffeomorphism. Prove that the formula

$$\Phi(x,\alpha) = (\varphi(x),((d\varphi)_x^{-1})^*\alpha)$$

defines a diffeomorphism of T^*L into itself. Determine $\Phi^*\lambda$ and prove that Φ preserves the symplectic from.

Exercise II.3. Let ω be a non degenerate 2-form on a manifold W. Define the Hamiltonian vector fields and Poisson brackets as above (this does not use the fact that ω is closed). Express

$$(d\omega)_x(X,Y,Z)$$

when X, Y et Z are tangent vectors to W at x that are the values at x of the Hamiltonian vector fields of three functions f, g and h. Prove that ω is a closed form if and only if the Poisson bracket it defines satisfies the Jacobi identity.

Exercise II.4. Assume X and Y are two "locally Hamiltonian" vector fields on a symplectic manifold, namely that $\iota_X\omega$ et $\iota_Y\omega$ are closed forms. Prove that their Lie bracket $[X,Y]$ is a globally Hamiltonian vector field, namely that $\iota_{[X,Y]}\omega$ is an *exact* form.

Exercise II.5 *(The Hodge star operator)*. Let V be an n-dimensional oriented manifold endowed with a Riemannian metric g and let α be the Riemannian volume form. Check that the formula

$$g(u_1 \wedge \cdots \wedge u_p, v_1 \wedge \cdots \wedge v_p) = \det(g(u_i,v_j)_{1 \le i,j \le p})$$

defines an metric on $\Lambda^p T^*V$... and that the map

$$\star : \Lambda^p T^*V \longrightarrow \Lambda^{n-p} T^*V$$

defined by

$$u \wedge (\star v) = g(u,v)\alpha$$

for all $u \in \Lambda^p T^*V$ defines, indeed, an operator, the *Hodge star operator*, which is an isometry. Check that

$$\star\star = (-1)^{p(n-1)} \operatorname{Id}_{\Lambda^p T^*V}.$$

Exercise II.6 *(Multilinear algebra in* $\mathbf{R}^4$ *).* Consider the vector space $\mathbf{R}^4$, with its Euclidean structure $g(X, Y) = (X, Y)$ and canonical basis (e_1, e_2, e_3, e_4), and the vector space

$$\Lambda = \Lambda^2(\mathbf{R}^4)^\star$$

of alternated bilinear forms on $\mathbf{R}^4$.

(1) What is the dimension of Λ? Check that Λ is isomorphic to the vector space of skew-symmetric endomorphisms of $\mathbf{R}^4$.

(2) Endow Λ with the Euclidean structure $(\ ,\ )$ induced by that of $\mathbf{R}^4$, namely such that the basis $(e_i^\star \wedge e_j^\star)/\sqrt{2}$ (for $1 \leq i < j \leq 4$) is orthonormal. Define the (Hodge) star operator $\star$ on Λ by the formula

$$(\star\alpha) \wedge \eta = (\alpha, \eta)\det \text{ for all } \eta \in \Lambda$$

(where det, namely the determinant, is the generator of $\Lambda^4(\mathbf{R}^4)^\star$ such that $\det(e_1 \wedge e_2 \wedge e_3 \wedge e_4) = 1$.

Check that $\star$ is an involution. Determine the $\star(e_i^\star \wedge e_j^\star)$ and the eigenspaces of $\star$.

(3) Call Λ_+ the subspace of forms that are invariant by $\star$ (they are called "self-dual" forms). To any α in Λ_+, associate as in (1) a skew-symmetric endomorphism

$$J_\alpha : \mathbf{R}^4 \longrightarrow \mathbf{R}^4.$$

Prove that $J_\alpha^2 = -\operatorname{Id}$ if and only if $(\alpha, \alpha) = 1$.

Exercise II.7 *(Calabi-Yau surfaces).* Let M be a Calabi-Yau surface with Kähler form ω and holomorphic 2-form Ω.

(1) There exists a local basis (φ_1, φ_2) of the vector space of holomorphic forms on M in which

$$\omega = \frac{1}{2i} (\varphi_1 \wedge \bar{\varphi}_1 + \varphi_2 \wedge \bar{\varphi}_2)$$

(see [15]). Prove that, on the open set where φ_1 and φ_2 are defined, one has

$$\Omega = \lambda \varphi_1 \wedge \varphi_2$$

for some *constant* λ.

(2) Check that $\star\Omega = \bar{\Omega}$. Deduce that the real α and imaginary β parts of Ω are self-dual in the sense that $\star\alpha = \alpha$ and $\star\beta = \beta$.

(3) Prove that the formula

$$\alpha(X, JY) = g(X, Y)$$

defines a skew-symmetric endomorphism J of the tangent bundle TM and that

$$JX \in \langle X, IX \rangle^\perp = (\mathbf{C} \cdot X)^\perp.$$

(4) Prove that $J^2 = -\mathrm{Id}$, so that J is an almost[6] complex structure on M, and that J is an isometry for g.

(5) Prove that M is endowed with a hyperkähler structure, namely with three isometries I, J and K that are almost complex structures and anti-commute and with three non degenerate 2-forms that are Kähler for the metric g and respectively for each of the complex structures I, J and K.

Exercise II.8. Using the notation of Exercise II.7, prove that the special Lagrangian submanifolds of the Calabi-Yau manifold M are the complex curves for the complex structure J.

Exercise II.9. In this exercise, W denotes a complex analytic manifold[7] of complex dimension 2, endowed with the structure of a Calabi-Yau manifold, with the Kähler form ω, the holomorphic volume form Ω and the metric γ. Assume moreover that $H^1_{DR}(W) = 0$.

Consider a vector field X on W, assume that it is not identically zero, and that it preserves ω and Ω, namely that it satisfies the relations $\mathcal{L}_X \omega = 0$ and $\mathcal{L}_X \Omega = 0$. Assume moreover that the 1-form $\iota_X \Omega$ is *holomorphic*.

The metric γ and the vector field X are assumed to be complete.

(1) Prove that X is the Hamiltonian vector field of a function $H : W \to \mathbf{R}$.

(2) Prove that $\iota_X \Omega$ is preserved by X and that there exists a holomorphic function $f : W \to \mathbf{C}$ such that $\iota_X \Omega = df$. Let $x \in W$ be a point such that $X_x \neq 0$. Prove that the kernel of $(\iota_X \Omega)_x$ is the complex line in $T_x W$ spanned by X_x.

(3) Assume that L is a Lagrangian submanifold of W that is preserved by X (this means that $X_x \in T_x L$ for all $x \in L$). Check that the connected components of L are contained in the level sets $H^{-1}(a)$ of the Hamiltonian H.

(4) Call g and h respectively the real and imaginary part of f. Assume now that L is a *special* Lagrangian submanifold. Prove that h is locally constant on L.

(5) Let $a \in \mathbf{R}$ be a *regular* value of H and let $Q = H^{-1}(a)$ be the corresponding level set in W. Fix a point x in Q. Prove that the orthogonal of X_x for the metric γ in $T_x Q$ is a complex line D_x and that the complex linear form $(\iota_X \Omega)_x$ is non zero on D_x. Deduce that the two real linear forms $dh(x)$ and $dH(x)$ are independent. Prove that, for all $b \in \mathbf{R}$, $L = Q \cap h^{-1}(b)$ is a dimension-2 submanifold of W and that it is special Lagrangian.

(6) Prove that $g\,|_L$ has no critical point.

---

[6] It is not very hard to prove that this is a *genuine* complex structure, namely that M is a complex manifold for some structure such that the multiplication by i is J.

[7] This exercise (slightly) generalizes the construction given in §I.5.e and in particular that of Exercise I.19.

(7) Assume that the Hamiltonian vector field X is periodic. Prove that the connected components of L are diffeomorphic to $S^1 \times \mathbf{R}$.

Bibliography

[1] V. I. ARNOLD – "A characteristic class entering in quantization conditions", *Funct. Anal. Appl.* **1** (1965).

[2] — , *Mathematical methods in classical mechanics*, Springer, 1978.

[3] — , "Lagrange and Legendre cobordisms I and II", *Funct. Anal. Appl.* **14** (1980), p. 167–177 et 252–260.

[4] V. I. ARNOLD et A. B. GIVENTAL – "Symplectic geometry", *Dynamical systems, Encyclopædia of Math. Sci., Springer* (1985).

[5] M. AUDIN – "Symplectic and almost complex manifolds", *in [7]*, p. 41–74.

[6] — , *Les systèmes hamiltoniens et leur intégrabilité*, Cours Spécialisés, 8, Société Mathématique de France & EDP Sciences, 2001.

[7] M. AUDIN et J. LAFONTAINE (éds.) – *Holomorphic curves in symplectic geometry*, Progress in Math., Birkhäuser, 1994.

[8] J. BERTIN, J.-P. DEMAILLY, L. ILLUSIE et C. PETERS – *Introduction à la théorie de Hodge*, Panoramas et Synthèses, 3, Société Mathématique de France, 1996.

[9] R. BRYANT – "Some examples of special Lagrangian tori", *preprint* (1998).

[10] H. CARTAN – *Calcul différentiel*, Hermann, Paris, 1967.

[11] S. DONALDSON – "Moment maps and diffeomorphisms", *Asian Journal of Math.* (2000), à paraître.

[12] D. B. FUKS – "Maslov-Arnold characteristic classes", *Soviet Math. Dokl.* **9** (1968), p. 96–99.

[13] A. B. GIVENTAL – "Lagrangian imbeddings of surfaces and the open Whitney umbrella", *Funktsional. Anal. i Prilozhen.* **20** (1986), no. 3, p. 35–41, 96.

[14] A. B. GIVENTAL – "Equivariant Gromov-Witten invariants", *Internat. Math. Res. Notices* **13** (1996), p. 613–663.

[15] P. A. GRIFFITHS et J. HARRIS – *Principles of algebraic geometry*, Wiley, 1978.

[16] M. GROMOV – "Pseudo-holomorphic curves in symplectic manifolds", *Invent. Math.* **82** (1985), p. 307–347.

[17] M. GROMOV – *Partial differential relations*, Springer, Berlin, 1986.

[18] R. HARVEY et H. B. LAWSON, JR. – "Calibrated geometries", *Acta Math.* **148** (1982), p. 47–157.

[19] N. J. HITCHIN – "The moduli space of special Lagrangian submanifolds", *Ann. Scuola Norm. Sup. Pisa Cl. Sci. (4)* **25** (1997), no. 3-4, p. 503–515 (1998), Dedicated to Ennio De Giorgi.

[20] — , "Lectures on special Lagrangian submanifolds", *preprint* (1999).

[21] V. M. KHARLAMOV – "On the classification of nonsingular surfaces of degree 4 in $\mathbf{RP}^3$ with respect to rigid isotopies", *Funktsional. Anal. i Prilozhen.* **18** (1984), no. 1, p. 49–56.

[22] J. LAFONTAINE – *Introduction aux variétés différentielles*, Presses universitaires de Grenoble, 1996.

[23] J. A. LEES – "On the classification of Lagrange immersions", *Duke Math. J.* **43** (1976), no. 2, p. 217–224.

[24] D. MCDUFF et D. SALAMON – *Introduction to symplectic topology*, The Clarendon Press Oxford University Press, New York, 1995, Oxford Science Publications.

[25] R. C. MCLEAN – "Deformations of calibrated submanifolds", *Comm. Anal. Geom.* **6** (1998), no. 4, p. 705–747.

[26] J. MILNOR – *Morse theory*, Princeton University Press, 1963.

[27] K. MOHNKE – *preprint* (2001).

[28] J. MOSER – "On the volume elements on a manifold", *Trans. Amer. Math. Soc.* **120** (1965), p. 286–294.

[29] H. ROSENBERG et D. HOFFMAN – *Surfaces minimales et solutions de problèmes variationnels*, Société Mathématique de France, Paris, 1993.

[30] A. CANNAS DA SILVA – *Lectures on symplectic geometry*, Lecture Notes in Mathematics, Springer, 2001.

[31] M. STENZEL – "Ricci-flat metrics on the complexification of a compact rank one symmetric space", *Manuscripta Math.* **80** (1993), p. 151–163.

[32] A. STROMINGER, S. T. YAU et E. ZASLOW – "Mirror symmetry is *T*-duality", *Nuclear Phys.* **B 479** (1996), p. 243–259.

[33] C. VOISIN – *Symétrie miroir*, Panoramas et Synthèses, 2, Société Mathématique de France, 1996, English translation: *Mirror symmetry*, SMF/AMS Texts and Monographs 1, 1999.

[34] A. WEINSTEIN – *Lectures on symplectic manifolds*, CBMS Regional Conference Series in Mathematics, 29, Amer. Math. Soc., 1977.

[35] S. T. YAU – "On the Ricci curvature of a compact Kähler manifold and the complex Monge-Ampère equation I", *Comm. Pure and Appl. Math.* **31** (1978), p. 339–411.

Part B

Symplectic Toric Manifolds

Ana Cannas da Silva

Foreword

The goal of this text is to provide a fast elementary introduction to toric manifolds (i.e., smooth toric varieties) from the symplectic viewpoint. The study of toric manifolds has many different entrances and has been scoring a wide spectrum of applications. For symplectic geometers, they provide examples of extremely symmetric and completely integrable hamiltonian spaces. In order to distinguish the algebraic from the symplectic approach, we call *algebraic toric manifolds* to the smooth toric varieties in algebraic geometry, and say *symplectic toric manifolds* when studying their symplectic properties.

Native to algebraic geometry, the theory of toric varieties has been around for about thirty years. It was introduced by Demazure in [16] who used toric varieties for classifying some algebraic subgroups. Since 1970 many nice surveys of the theory of toric varieties have appeared (see, for instance, [14, 21, 28, 41]). Algebraic geometers and combinatorialists have found fruitful applications of toric varieties to the geometry of convex polytopes, resolutions of singularities, compactifications of locally symmetric spaces, critical points of analytic functions, etc. For the last ten years, toric geometry became an important tool in physics in connection with mirror symmetry [13] where research has been intensive.

In this text we emphasize the geometry of the *moment map* whose image, the so-called *moment polytope*, determines the symplectic toric manifolds. The notion of a moment map associated to a group action generalizes that of a hamiltonian function associated to a vector field. Either of these notions formalizes the Noether principle, which states that to every symmetry (such as a group action) in a mechanical system, there corresponds a conserved quantity. The concept of a moment map was introduced by Souriau [45] under the french name *application moment*; besides the more standard english translation to *moment map*, the alternative *momentum map* is also used. Moment maps have been asserting themselves as a main tool to study problems in geometry and topology when there is a suitable symmetry, as illustrated in the book by Gelfand, Kapranov and Zelevinsky [22]. The material in some sections of the second part of these notes borrows largely from that excellent text, where details are given and where the discussion continues in exciting new directions.

This course splits into two parts: the first concentrates on the *symplectic viewpoint*, and the second focuses on the *algebraic viewpoint* with links to symplectic geometry. Each of the two parts has a similar structure: introduction, classification, polytopes. So the lectures seem periodic though the languages for each half are noticeably different.

After introducing, in the first lecture, basic notions related to symplectic toric manifolds, in Lecture 2 we state their classification, and prove the existence part of the classification theorem by using the technique of symplectic reduction. Lecture 3 discusses moment polytopes, namely how to read from the polytope some topological properties of the corresponding symplectic toric manifolds, as well as how are some changes on the polytopes translated into changes of the corresponding symplectic toric manifolds. Lecture 4 introduces toric manifolds from the algebro-geometric perspective, after reviewing definitions and notation in the theory of algebraic varieties. Lecture 5 describes the classification of toric varieties using the language of spectra and fans. Finally, Lecture 6 deals with polytopes now from the algebraic point of view, studying some geometric properties of toric varieties which polytopes encode. Lectures 3 and 6 underline the moment map potential.

Geometry of manifolds at the level of a first-year graduate course is the basic prerequisite for this course. Some familiarity with symplectic geometry is useful to read these notes faster, though most of the needed definitions and results are stated here. Scattered through the text, there are exercises designed to complement the exposition or extend the reader's understanding. Throughout, the symbol $\mathbb{P}^n$ denotes n-(complex-)dimensional complex projective space.

Chapter I

Symplectic Viewpoint

I.1 Symplectic Toric Manifolds

In order to define symplectic toric manifolds, we begin by introducing the basic objects in symplectic/hamiltonian geometry/mechanics which lead to their consideration. Our discussion centers around moment maps.

I.1.1 Symplectic Manifolds

Definition I.1.1. *A* **symplectic form** *on a manifold M is a closed 2-form on M which is nondegenerate at every point of M. A* **symplectic manifold** *is a pair (M, ω) where M is a manifold and ω is a symplectic form on M.*

By linear algebra, a symplectic manifold is necessarily even-dimensional.

Examples.

1. Let $M = \mathbb{R}^{2n}$ with linear coordinates $x_1, \ldots, x_n, y_1, \ldots, y_n$. The **standard symplectic form on $\mathbb{R}^{2n}$** is

$$\omega_0 = \sum_{k=1}^{n} dx_k \wedge dy_k \ .$$

2. Let $M = \mathbb{C}^n$ with linear coordinates $z_1, \ldots, z_n$. The form

$$\omega_0 = \frac{i}{2} \sum_{k=1}^{n} dz_k \wedge d\bar{z}_k$$

is a symplectic form on $\mathbb{C}^n$. In fact, this form equals that of the previous example under the identification $\mathbb{C}^n \simeq \mathbb{R}^{2n}$, $z_k = x_k + iy_k$.

3. Let $M = S^2$ regarded as the set of unit vectors in $\mathbb{R}^3$. Tangent vectors to S^2 at p may then be identified with vectors orthogonal to p. The **standard symplectic form on** S^2 is the form induced by the inner and exterior products:

$$\omega_p(u, v) := \langle p, u \times v \rangle , \qquad \text{for } u, v \in T_p S^2 = \{p\}^\perp .$$

This form is closed because it is of top degree; it is nondegenerate because $\langle p, u \times v \rangle \neq 0$ when $u \neq 0$ and we take, for instance, $v = u \times p$.

Exercise 1

Check that, in cylindrical coordinates away from the poles ($0 \leq \theta < 2\pi$ and $-1 < h < 1$), the standard symplectic form on S^2 is the area form given by

$$\omega_{\text{standard}} = d\theta \wedge dh .$$

The natural notion of equivalence in the symplectic category is expressed by a *symplectomorphism*:

Definition I.1.2. *Let (M_1, ω_1) and (M_2, ω_2) be $2n$-dimensional symplectic manifolds, and let $\varphi : M_1 \to M_2$ be a diffeomorphism. Then φ is a* **symplectomorphism** *if $\varphi^* \omega_2 = \omega_1$.*

The Darboux theorem (see, for instance, [12] or Theorem I.3.1 for the case where the group is trivial) states that any symplectic manifold (M^{2n}, ω) is locally symplectomorphic to $(\mathbb{R}^{2n}, \omega_0)$. In other words, the prototype of a local piece of a $2n$-dimensional symplectic manifold is $(\mathbb{R}^{2n}, \omega_0)$. Hence, this theorem provides the local classification of symplectic manifolds in terms of a unique invariant: the dimension.

Let (M, ω) be a symplectic manifold of dimension $2n$. A **Darboux chart** for M is a chart $(\mathcal{U}, x_1, \ldots, x_n, y_1, \ldots, y_n)$ such that

$$\omega|_{\mathcal{U}} = \sum_{k=1}^{n} dx_k \wedge dy_k .$$

By the Darboux theorem, there exists a Darboux chart centered at each point of a symplectic manifold.

I.1.2   Hamiltonian Vector Fields

Let (M, ω) be a symplectic manifold.

Definition I.1.3. *A vector field X on M is* **symplectic** *if the contraction $\imath_X \omega$ is closed. A vector field X on M is* **hamiltonian** *if the contraction $\imath_X \omega$ is exact.*

Locally on every contractible open set, every symplectic vector field is hamiltonian. If the first de Rham cohomology group is trivial, then globally every symplectic vector field is hamiltonian; in general, $H^1_{\mathrm{deRham}}(M)$ measures the obstruction for symplectic vector fields to be hamiltonian.

Note that the flow of a symplectic vector field X preserves the symplectic form:

$$\mathcal{L}_X \omega = d \underbrace{\imath_X \omega}_{\text{closed}} + \imath_X \underbrace{d\omega}_{0} = 0 \ .$$

If a vector field X is hamiltonian with $\imath_X \omega = dH$ for some smooth function $H : M \to \mathbb{R}$, then the flow of X also preserves the function H:

$$\mathcal{L}_X H = \imath_X dH = \imath_X \imath_X \omega = 0 \ .$$

Therefore, each integral curve $\{\rho_t(x) \mid t \in \mathbb{R}\}$ of X must be contained in a level set of H:

$$H(x) = (\rho_t^* H)(x) = H(\rho_t(x)) \ , \quad \forall t \ .$$

Definition I.1.4. *A **hamiltonian function** for a hamiltonian vector field X on M is a smooth function $H : M \to \mathbb{R}$ such that $\imath_X \omega = dH$.*

By nondegeneracy of ω, any function $H \in C^\infty(M)$ is a hamiltonian function for some hamiltonian vector field because the equation $\imath_X \omega = dH$ can be always solved for a smooth vector field X. A hamiltonian vector field X defines a hamiltonian function *up to* a locally constant function.

Examples.

1. On the standard symplectic 2-sphere $(S^2, d\theta \wedge dh)$, the vector field $X = \frac{\partial}{\partial \theta}$ is hamiltonian with hamiltonian function given by the height function:

$$\imath_X (d\theta \wedge dh) = dh \ .$$

   The motion generated by this vector field is rotation about the vertical axis, which of course preserves both area and height.

2. On the symplectic 2-torus $(\mathbb{T}^2, d\theta_1 \wedge d\theta_2)$, the vector fields $X_1 = \frac{\partial}{\partial \theta_1}$ and $X_2 = \frac{\partial}{\partial \theta_2}$ are symplectic but not hamiltonian.

I.1.3   Integrable Systems

Let X_H denote a hamiltonian vector field on a symplectic manifold (M, ω) with hamiltonian function $H \in C^\infty(M)$.

Definition I.1.5. *The* **Poisson bracket** *of two functions* $f, g \in C^\infty(M)$ *is the function*

$$\{f, g\} := \omega(X_f, X_g) \ .$$

We have $X_{\{f,g\}} = -[X_f, X_g]$ because $X_{\omega(X_f, X_g)} = [X_g, X_f]$, where $[\cdot, \cdot]$ is the Lie bracket of vector fields.[1]

> **Exercise 2**
> Check that the Poisson bracket $\{\cdot, \cdot\}$ is a Lie bracket and satisfies the **Leibniz rule**:
> $$\{f, gh\} = \{f, g\}h + g\{f, h\} \qquad \forall f, g, h \in C^\infty(M) \ .$$

Therefore, if (M, ω) is a symplectic manifold, then $(C^\infty(M), \{\cdot, \cdot\})$ is a Poisson algebra, that is a Lie algebra with an associative product for which the Leibniz rule holds. Furthermore, we have an anti-homomorphism of Lie algebras

$$
\begin{array}{ccc}
C^\infty(M) & \longrightarrow & \chi(M) \\
H & \longmapsto & X_H \ .
\end{array}
$$

Definition I.1.6. *A* **hamiltonian system** *is a triple* (M, ω, H), *where* (M, ω) *is a symplectic manifold and* $H \in C^\infty(M)$ *is a function, called the* **hamiltonian function**.

> **Exercise 3**
> Show that $\{f, H\} = 0$ if and only if f is constant along integral curves of X_H.

A function f as in the previous exercise is called an **integral of motion** (or a *first integral* or a *constant of motion*) for the hamiltonian system (M, ω, H).

In general, hamiltonian systems do not admit integrals of motion which are *independent* of the hamiltonian function. Functions $f_1, \ldots, f_n$ on M are said to be **independent** if their differentials $(df_1)_p, \ldots, (df_n)_p$ are linearly independent at all points p in some open dense subset of M. Loosely speaking, a hamiltonian system is *(completely) integrable* if it has as many commuting integrals of motion as possible. Commutativity is with respect to the Poisson bracket. Notice that, if $f_1, \ldots, f_n$ are commuting integrals of motion for a hamiltonian system (M, ω, H), then, at each $p \in M$, their hamiltonian vector fields generate an **isotropic** subspace of $T_p M$:

$$\omega(X_{f_i}, X_{f_j}) = \{f_i, f_j\} = 0 \ .$$

If $f_1, \ldots, f_n$ are independent at p, then, by symplectic linear algebra, n can be at most half the dimension of M.

---

[1]The Lie bracket of two vector fields is their commutator, where they are regarded as first-order differential operators. A bilinear function $\{\cdot, \cdot\} : C^\infty(M) \times C^\infty(M) \to C^\infty(M)$ is a **Lie bracket** if it is antisymmetric, i.e., $\{f, g\} = -\{g, f\}$, $\forall f, g \in C^\infty(M)$, and satisfies the Jacobi identity:

$$\{f, \{g, h\}\} + \{g, \{h, f\}\} + \{h, \{f, g\}\} = 0 \qquad \forall f, g, h \in C^\infty(M) \ .$$

Definition I.1.7. *A hamiltonian system* (M, ω, H) *is* (**completely**) **integrable** *if it admits* $n = \frac{1}{2} \dim M$ *independent integrals of motion,* $f_1 = H, f_2, \ldots, f_n$, *which are pairwise in involution with respect to the Poisson bracket, i.e.,* $\{f_i, f_j\} = 0$, *for all* i, j.

Example. A hamiltonian system (M, ω, H) where M is 4-dimensional is integrable if there is an integral of motion independent of H (the commutativity condition is automatically satisfied). $\diamond$

For interesting examples of integrable systems, see [7].

Let (M, ω, H) be an integrable system of dimension $2n$ with integrals of motion $f_1 = H, f_2, \ldots, f_n$. Let $c \in \mathbb{R}^n$ be a regular value of $f := (f_1, \ldots, f_n)$. The corresponding level set, $f^{-1}(c)$, is a *lagrangian submanifold*. A submanifold Y of a $2n$-dimensional symplectic manifold (M, ω) is **lagrangian** if it is n-dimensional and if $i^*\omega = 0$ where $i : Y \hookrightarrow M$ is the inclusion map.

> **Exercise 4**
> By following the flows, show that if the hamiltonian vector fields $X_{f_1}, \ldots, X_{f_n}$ are complete on the level $f^{-1}(c)$ (i.e., if their flows are defined for all time), the connected components of $f^{-1}(c)$ are homogeneous spaces for $\mathbb{R}^n$, i.e., are of the form $\mathbb{R}^{n-k} \times \mathbb{T}^k$ for some $0 \leq k \leq n$, where $\mathbb{T}^k$ is a k-dimensional torus.

Any compact component of $f^{-1}(c)$ must hence be a torus. These compact components, when they exist, are called **Liouville tori**.

Theorem I.1.8. (Arnold-Liouville [2]) *Let* (M, ω, H) *be an integrable system of dimension* $2n$ *with integrals of motion* $f_1 = H, f_2, \ldots, f_n$. *Let* $c \in \mathbb{R}^n$ *be a regular value of* $f := (f_1, \ldots, f_n)$. *The corresponding level* $f^{-1}(c)$ *is a lagrangian submanifold of* M.

(a) *If the flows of* $X_{f_1}, \ldots, X_{f_n}$ *starting at a point* $p \in f^{-1}(c)$ *are complete, then the connected component of* $f^{-1}(c)$ *containing* p *is a homogeneous space for* $\mathbb{R}^n$. *Namely, there is an affine structure on that component with coordinates* $\varphi_1, \ldots, \varphi_n$, *known as* **angle coordinates**, *in which the flows of the vector fields* $X_{f_1}, \ldots, X_{f_n}$ *are linear.*

(b) *There are coordinates* $\psi_1, \ldots, \psi_n$, *known as* **action coordinates**, *complementary to the angle coordinates such that the* ψ_i's *are integrals of motion and* $\varphi_1, \ldots, \varphi_n, \psi_1, \ldots, \psi_n$ *form a Darboux chart.*

Therefore, the dynamics of an integrable system is extremely simple and the system has an explicit solution in action-angle coordinates. The proof of part (a) – the easy, yet interesting, part – of the Arnold-Liouville theorem is sketched above. For the proof of part (b), see [2, 17].

Geometrically, part (a) of the Arnold-Liouville theorem says that, in a neighborhood of the value c, the map $f : M \to \mathbb{R}^n$ collecting the given integrals of motion is a **lagrangian fibration**, i.e., it is locally trivial and its fibers are lagrangian submanifolds. The coordinates along the fibers are the angle coordinates.[2] Part (b) of the theorem guarantees the existence of coordinates on $\mathbb{R}^n$, the action coordinates, which (Poisson) commute among themselves and which satisfy $\{\varphi_i, \psi_j\} = \delta_{ij}$ with respect to the angle coordinates. Notice that, in general, the action coordinates are not the given integrals of motion because $\varphi_1, \ldots, \varphi_n, f_1, \ldots, f_n$ do not form a Darboux chart.

I.1.4   Hamiltonian Actions

Definition I.1.9. *An* **action** *of a Lie group* G *on a manifold* M *is a group homomorphism*

$$\psi: \quad G \quad \longrightarrow \quad \mathrm{Diff}(M)$$
$$g \quad \longmapsto \quad \psi_g \;,$$

where $\mathrm{Diff}(M)$ *is the group of diffeomorphisms of* M. *The* **evaluation map** *associated with an action* $\psi: G \to \mathrm{Diff}(M)$ *is*

$$\mathrm{ev}_\psi: \quad M \times G \quad \longrightarrow \quad M$$
$$(p, g) \quad \longmapsto \quad \psi_g(p) \;.$$

The action ψ *is* **smooth** *if* ev_ψ *is a smooth map.*

We will always assume that an action is smooth.

Example. Complete vector fields[3] on a manifold M are in one-to-one correspondence with actions of $\mathbb{R}$ on M. The diffeomorphism $\psi_t : M \to M$ associated to $t \in \mathbb{R}$ is the time-t map $\exp tX$ defined by the flow of the vector field X.   ◇

Let (M, ω) be a symplectic manifold, and G a Lie group with an action $\psi : G \to \mathrm{Diff}(M)$.

Definition I.1.10. *The action* ψ *is a* **symplectic action** *if it is by symplectomorphisms, i.e.,*

$$\psi: G \longrightarrow \mathrm{Sympl}(M, \omega) \subset \mathrm{Diff}(M) \;,$$

where $\mathrm{Sympl}(M, \omega)$ *is the group of symplectomorphisms of* (M, ω).

Examples.

1. On the symplectic 2-sphere $(S^2, d\theta \wedge dh)$ in cylindrical coordinates, the one-parameter group of diffeomorphisms given by rotation around the vertical axis, $\psi_t(\theta, h) = (\theta + t, h)$ $(t \in \mathbb{R})$ is a symplectic action of the group $S^1 \simeq \mathbb{R}/\langle 2\pi \rangle$, as it preserves the area form $d\theta \wedge dh$.

---

[2]The name *angle coordinates* is used even if the fibers are not tori.
[3]A vector field is **complete** if its integral curves through each point exist for *all* time.

2. On the symplectic 2-torus $(\mathbb{T}^2, d\theta_1 \wedge d\theta_2)$, the one-parameter groups of diffeomorphisms given by rotation around each circle, $\psi_{1,t}(\theta_1, \theta_2) = (\theta_1 + t, \theta_2)$ $(t \in \mathbb{R})$ and $\psi_{2,t}$ similarly defined, are symplectic actions of S^1.

Let (M, ω) be a symplectic manifold, G a Lie group with an action $\psi : G \to$ Diff(M), and $\mathfrak{g}$ the Lie algebra of G with dual vector space $\mathfrak{g}^*$.

Definition I.1.11. *The action ψ is a* **hamiltonian action** *if there exists a map*

$$\mu : M \longrightarrow \mathfrak{g}^*$$

satisfying the following two conditions:

- *For each $X \in \mathfrak{g}$, let $\mu^X : M \to \mathbb{R}$, $\mu^X(p) := \langle \mu(p), X \rangle$, be the component of μ along X, and let $X^\#$ be the vector field on M generated by the one-parameter subgroup $\{\exp tX \mid t \in \mathbb{R}\} \subseteq G$. Then*

$$d\mu^X = \imath_{X^\#}\omega$$

i.e., the function μ^X is a hamiltonian function for the vector field $X^\#$.

- *The map μ is equivariant with respect to the given action ψ of G on M and the coadjoint action Ad^* of G on $\mathfrak{g}^*$:*

$$\mu \circ \psi_g = \mathrm{Ad}_g^* \circ \mu , \qquad \text{for all } g \in G .$$

The vector (M, ω, G, μ) is then called a **hamiltonian G-space** *and μ is called a* **moment map**.

Exercise 5
Check that complete symplectic vector fields on M are in one-to-one correspondence with symplectic actions of $\mathbb{R}$ on M, and that, similarly, complete hamiltonian vector fields on M are in one-to-one correspondence with hamiltonian actions of $\mathbb{R}$ on M.

Examples. Consider the previous set of two examples (as well as in the examples of Section I.1.2). The first – regarding S^2 – is an example of a hamiltonian action of S^1 with moment map given by the height function, under a suitable identification of the dual of the Lie algebra of S^1 with $\mathbb{R}$. The second example – regarding T^2 – is not hamiltonian since the one-forms $d\theta_1$ and $d\theta_2$ are not exact. ◇

Exercise 6
Let G be any compact Lie group and H a closed subgroup of G, with $\mathfrak{g}$ and $\mathfrak{h}$ the respective Lie algebras. The projection $i^* : \mathfrak{g}^* \to \mathfrak{h}^*$ is the map dual to the inclusion $i : \mathfrak{h} \hookrightarrow \mathfrak{g}$. Suppose that (M, ω, G, ϕ) is a hamiltonian G-space. Show that the restriction of the G-action to H is hamiltonian with moment map

$$i^* \circ \phi : M \longrightarrow \mathfrak{h}^* .$$

Exercise 7
Suppose that a Lie group G acts in a hamiltonian way on two symplectic manifolds (M_j, ω_j), $j = 1, 2$, with moment maps $\mu_j : M_j \to \mathfrak{g}^*$. The product manifold $M_1 \times M_2$ has a natural product symplectic structure given by the sum of the pull-backs of the symplectic forms on each factor, via the two projections. Prove that the diagonal action of G on $M_1 \times M_2$ is hamiltonian with moment map $\mu : M_1 \times M_2 \to \mathfrak{g}^*$ given by

$$\mu(p_1, p_2) = \mu_1(p_1) + \mu_2(p_2) , \quad \text{for } p_j \in M_j .$$

From now on, we concentrate on actions of a torus $G = \mathbb{T}^m = \mathbb{R}^m/\mathbb{Z}^m$.

I.1.5   Hamiltonian Torus Actions

The coadjoint action is trivial on a torus (a product of circles $S^1 \times \cdots \times S^1$). Hence, if $G = \mathbb{T}^n$ is an n-dimensional torus with Lie algebra and its dual both identified with euclidean space, $\mathfrak{g} \simeq \mathbb{R}^n$ and $\mathfrak{g}^* \simeq \mathbb{R}^n$, a moment map for an action of G on (M, ω) is simply a map $\mu : M \longrightarrow \mathbb{R}^n$ satisfying:

- *For each basis vector X_i of $\mathbb{R}^n$, the function μ^{X_i} is a hamiltonian function for $X_i^\#$ and is invariant under the action of the torus.*

If $\mu : M \to \mathbb{R}^n$ is a moment map for a torus action, then clearly any of its translations $\mu + c$ ($c \in \mathbb{R}^n$) is also a moment map for that action. Reciprocally, any two moment maps for a given hamiltonian torus action differ by a constant.

Example. On $(\mathbb{C}, \omega_0 = \frac{i}{2} dz \wedge d\bar{z})$, consider the action of the circle $S^1 = \{t \in \mathbb{C} : |t| = 1\}$ by rotations

$$\psi_t(z) = t^k z , \qquad t \in S^1 ,$$

where $k \in \mathbb{Z}$ is fixed. The action $\psi : S^1 \to \text{Diff}(\mathbb{C})$ is hamiltonian with moment map $\mu : \mathbb{C} \to \mathfrak{g}^* \simeq \mathbb{R}$ given by

$$\mu(z) = -\tfrac{1}{2} k |z|^2 .$$

This can be easily checked in polar coordinates, since $\omega_0 = r\, dr \wedge d\theta$, $\mu(re^{i\theta}) = -\frac{1}{2} kr^2$ and the vector field on $\mathbb{C}$ corresponding to the generator 1 of $\mathfrak{g} \simeq \mathbb{R}$ is $X^\# = k \frac{\partial}{\partial \theta}$. $\diamond$

Exercise 8
Let $\mathbb{T}^n = \{(t_1, \ldots, t_n) \in \mathbb{C}^n : |t_j| = 1, \text{ for all } j\}$ be a torus acting diagonally on $\mathbb{C}^n$ by

$$(t_1, \ldots, t_n) \cdot (z_1, \ldots, z_n) = (t_1^{k_1} z_1, \ldots, t_n^{k_n} z_n) ,$$

where $k_1, \ldots, k_n \in \mathbb{Z}$ are fixed. Check that this action is hamiltonian with moment map $\mu : \mathbb{C}^n \to \mathfrak{g}^* \simeq \mathbb{R}^n$ given by

$$\mu(z_1, \ldots, z_n) = -\tfrac{1}{2}(k_1 |z_1|^2, \ldots, k_n |z_n|^2) \; (+ \text{constant}) .$$

Theorem I.1.12. (Atiyah [3], Guillemin-Sternberg [25]) *Let (M, ω) be a compact connected symplectic manifold, and let $\mathbb{T}^m$ be an m-torus. Suppose that $\psi : \mathbb{T}^m \to \mathrm{Sympl}(M, \omega)$ is a hamiltonian action with moment map $\mu : M \to \mathbb{R}^m$. Then:*

(a) *the levels of μ are connected;*

(b) *the image of μ is convex;*

(c) *the image of μ is the convex hull of the images of the fixed points of the action.*

The image $\mu(M)$ of the moment map is called the **moment polytope**. A proof of Theorem I.1.12 can be found in [36].

An action of a group G on a manifold M is called **effective** if it is injective as a map $G \to \mathrm{Diff}(M)$, i.e., each group element $g \neq e$ moves at least one point, that is, $\cap_{p \in M} G_p = \{e\}$, where $G_p = \{g \in G \mid g \cdot p = p\}$ is the stabilizer of p.

Exercise 9
Suppose that $\mathbb{T}^m$ acts linearly on $(\mathbb{C}^n, \omega_0)$. Let $\lambda^{(1)}, \ldots, \lambda^{(n)} \in \mathbb{Z}^m$ be the weights appearing in the corresponding weight space decomposition (further discussed in Section II.3.2), that is,

$$\mathbb{C}^n \simeq \bigoplus_{k=1}^{n} V_{\lambda^{(k)}} ,$$

where, for $\lambda^{(k)} = (\lambda_1^{(k)}, \ldots, \lambda_m^{(k)})$, $\mathbb{T}^m$ acts on the complex line $V_{\lambda^{(k)}}$ by

$$(e^{it_1}, \ldots, e^{it_m}) \cdot v = e^{i \sum_j \lambda_j^{(k)} t_j} v , \qquad \forall v \in V_{\lambda^{(k)}} , \ \forall k = 1, \ldots, n .$$

(a) Show that, if the action is effective, then $m \leq n$ and the weights $\lambda^{(1)}, \ldots, \lambda^{(n)}$ are part of a $\mathbb{Z}$-basis of $\mathbb{Z}^m$.

(b) Show that, if the action is symplectic (hence, hamiltonian), then the weight spaces $V_{\lambda^{(k)}}$ are symplectic subspaces.

(c) Show that, if the action is hamiltonian, then a moment map is given by

$$\mu(v) = -\frac{1}{2} \sum_{k=1}^{n} \lambda^{(k)} ||v_{\lambda^{(k)}}||^2 \ (\ + \text{ constant }) ,$$

where $|| \cdot ||$ is the standard norm[a] and $v = v_{\lambda^{(1)}} + \ldots + v_{\lambda^{(n)}}$ is the weight space decomposition. Cf. Exercise 8.

(d) Conclude that, if $\mathbb{T}^n$ acts on $\mathbb{C}^n$ in a linear, effective and hamiltonian way, then any moment map μ is a submersion, i.e., each differential $d\mu_v : \mathbb{C}^n \to \mathbb{R}^n$ ($v \in \mathbb{C}^n$) is surjective.

---

[a]Notice that the standard inner product satisfies $(v, w) = \omega_0(v, Jv)$ where $J \frac{\partial}{\partial z} = i \frac{\partial}{\partial z}$ and $J \frac{\partial}{\partial \bar{z}} = -i \frac{\partial}{\partial \bar{z}}$. In particular, the standard norm is invariant for a symplectic complex-linear action.

The following two results use the crucial fact that any effective action $\mathbb{T}^m \to \mathrm{Diff}(M)$ has orbits of dimension m; a proof may be found in [10].

Corollary I.1.13. *Under the conditions of the convexity theorem, if the $\mathbb{T}^m$-action is effective, then there must be at least $m+1$ fixed points.*

Proof. At a point p of an m-dimensional orbit the moment map is a submersion, i.e., $(d\mu_1)_p, \ldots, (d\mu_m)_p$ are linearly independent. Hence, $\mu(p)$ is an interior point of $\mu(M)$, and $\mu(M)$ is a nondegenerate convex polytope. Any nondegenerate convex polytope in $\mathbb{R}^m$ must have at least $m+1$ vertices. The vertices of $\mu(M)$ are images of fixed points. $\square$

Theorem I.1.14. *Let $(M, \omega, \mathbb{T}^m, \mu)$ be a hamiltonian $\mathbb{T}^m$-space. If the $\mathbb{T}^m$-action is effective, then $\dim M \geq 2m$.*

Proof. Since the moment map is constant on an orbit $\mathcal{O}$, for $p \in \mathcal{O}$ the exterior derivative

$$d\mu_p : T_pM \longrightarrow \mathfrak{g}^*$$

maps $T_p\mathcal{O}$ to 0. Thus

$$T_p\mathcal{O} \subseteq \ker d\mu_p = (T_p\mathcal{O})^\omega \ ,$$

where $(T_p\mathcal{O})^\omega$ is the symplectic orthogonal of $T_p\mathcal{O}$. This shows that orbits $\mathcal{O}$ of a hamiltonian torus action are always isotropic submanifolds of M. In particular, by symplectic linear algebra we have that $\dim \mathcal{O} \leq \frac{1}{2} \dim M$. Now consider an m-dimensional orbit. $\square$

I.1.6 Symplectic Toric Manifolds

Definition I.1.15. *A* **symplectic toric manifold** *is a compact connected symplectic manifold (M, ω) equipped with an effective hamiltonian action of a torus $\mathbb{T}$ of dimension equal to half the dimension of the manifold,*

$$\dim \mathbb{T} = \frac{1}{2} \dim M \ ,$$

and with a choice of a corresponding moment map μ.

> **Exercise 10**
> Show that an effective hamiltonian action of a torus $\mathbb{T}^n$ on a $2n$-dimensional symplectic manifold gives rise to an integrable system.
>
> **Hint:** The coordinates of the moment map are commuting integrals of motion.

Definition I.1.16. *Two symplectic toric manifolds, $(M_i, \omega_i, \mathbb{T}_i, \mu_i)$, $i = 1, 2$, are* **equivalent** *if there exists an isomorphism $\lambda : \mathbb{T}_1 \to \mathbb{T}_2$ and a λ-equivariant symplectomorphism $\varphi : M_1 \to M_2$ such that $\mu_1 = \mu_2 \circ \varphi$.*

Equivalent symplectic toric manifolds are often undistinguished.

Examples of symplectic toric manifolds.

1. The circle S^1 acts on the 2-sphere $(S^2, \omega_{\text{standard}} = d\theta \wedge dh)$ by rotations

$$e^{it} \cdot (\theta, h) = (\theta + t, h)$$

with moment map $\mu = h$ equal to the height function and moment polytope $[-1, 1]$.

Equivalently, the circle S^1 acts on $\mathbb{P}^1 = \mathbb{C}^2 - 0/\sim$ with the Fubini-Study form $\omega_{\text{FS}} = \frac{1}{4}\omega_{\text{standard}}$, by $e^{it} \cdot [z_0 : z_1] = [z_0 : e^{it} z_1]$. This is hamiltonian with moment map $\mu[z_0 : z_1] = -\frac{1}{2} \cdot \frac{|z_1|^2}{|z_0|^2 + |z_1|^2}$, and moment polytope $\left[-\frac{1}{2}, 0\right]$.

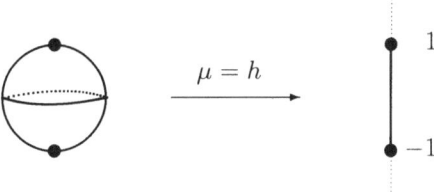

2. Let $(\mathbb{P}^2, \omega_{\text{FS}})$ be 2-(complex-)dimensional complex projective space equipped with the Fubini-Study form defined in Section I.2.3. The $\mathbb{T}^2$-action on $\mathbb{P}^2$ by $(e^{i\theta_1}, e^{i\theta_2}) \cdot [z_0 : z_1 : z_2] = [z_0 : e^{i\theta_1} z_1 : e^{i\theta_2} z_2]$ has moment map

$$\mu[z_0 : z_1 : z_2] = -\frac{1}{2} \left(\frac{|z_1|^2}{|z_0|^2 + |z_1|^2 + |z_2|^2}, \frac{|z_2|^2}{|z_0|^2 + |z_1|^2 + |z_2|^2} \right).$$

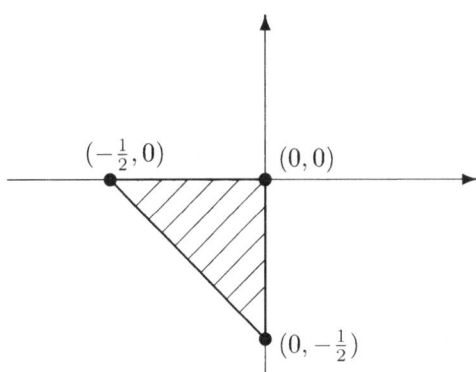

The fixed points get mapped as

$$\begin{array}{rcl}
[1:0:0] & \longmapsto & (0,0) \\
[0:1:0] & \longmapsto & \left(-\tfrac{1}{2},0\right) \\
[0:0:1] & \longmapsto & \left(0,-\tfrac{1}{2}\right)
\end{array}$$

Notice that the stabilizer of a preimage of the edges is S^1, while the action is free at preimages of interior points of the moment polytope.

Exercise 11
Compute a moment polytope for the $\mathbb{T}^3$-action on $\mathbb{P}^3$ as

$$(e^{i\theta_1}, e^{i\theta_2}, e^{i\theta_3}) \cdot [z_0 : z_1 : z_2 : z_3] = [z_0 : e^{i\theta_1} z_1 : e^{i\theta_2} z_2 : e^{i\theta_3} z_3] \ .$$

Exercise 12
Compute a moment polytope for the $\mathbb{T}^2$-action on $\mathbb{P}^1 \times \mathbb{P}^1$ as

$$(e^{i\theta}, e^{i\eta}) \cdot ([z_0 : z_1], [w_0 : w_1]) = ([z_0 : e^{i\theta} z_1], [w_0 : e^{i\eta} w_1]) \ .$$

I.2 Classification

Recall that a $2n$-dimensional symplectic toric manifold is a compact connected symplectic manifold (M^{2n}, ω) equipped with an effective hamiltonian action of an n-torus $\mathbb{T}^n$ and with a corresponding moment map $\mu : M \to \mathbb{R}^n$. In this lecture we describe the classification of equivalence classes of symplectic toric manifolds by their moment polytopes $\mu(M)$. Symplectic reduction is the quotienting technique which we use for the construction of a symplectic toric manifold out of an appropriate polytope, thus proving the existence part in the classification theorem.

I.2.1 Delzant's Theorem

We now define the class of polytopes[4] which arise in the classification of symplectic toric manifolds.

Definition I.2.1. *A* **Delzant polytope** Δ *in* $\mathbb{R}^n$ *is a polytope satisfying:*

- **simplicity**, *i.e., there are* n *edges meeting at each vertex;*

- **rationality**, *i.e., the edges meeting at the vertex* p *are rational in the sense that each edge is of the form* $p + tu_i$, $t \geq 0$, *where* $u_i \in \mathbb{Z}^n$;

- **smoothness**, *i.e., for each vertex, the corresponding* $u_1, \dots, u_n$ *can be chosen to be a* $\mathbb{Z}$-*basis of* $\mathbb{Z}^n$.

Examples of Delzant polytopes in $\mathbb{R}^2$:

The dotted vertical line in the trapezoidal example is there just to stress that it is a picture of a rectangle plus an isosceles triangle. For "taller" triangles, smoothness would be violated. "Wider" triangles (with integral slope) may still be Delzant. The family of the Delzant trapezoids of this type, starting with the rectangle, correspond, under the Delzant construction, to *Hirzebruch surfaces*; see Lecture 3.

$\diamond$

---

[4]A **polytope** in $\mathbb{R}^n$ is the convex hull of a finite number of points in $\mathbb{R}^n$. A **convex polyhedron** is a subset of $\mathbb{R}^n$ which is the intersection of a finite number of affine half-spaces. Hence, polytopes coincide with bounded convex polyhedra.

Examples of polytopes which are not Delzant:

The picture on the left fails the smoothness condition, since the triangle is not isosceles, whereas the one on the right fails the simplicity condition.   ◇

Delzant's theorem classifies (equivalence classes of) symplectic toric manifolds in terms of the combinatorial data encoded by a Delzant polytope.

Theorem I.2.2. (Delzant [15]) *Toric manifolds are classified by Delzant polytopes. More specifically, the bijective correspondence between these two sets is given by the moment map:*

$$\{toric\ manifolds\} \overset{1-1}{\longrightarrow} \{Delzant\ polytopes\}$$
$$(M^{2n}, \omega, \mathbb{T}^n, \mu) \longmapsto \mu(M)\ .$$

In Section I.2.5, we describe the construction which proves the (easier) existence part, or surjectivity, in Delzant's theorem. In order to prepare that, we will next give an algebraic description of Delzant polytopes.

Let Δ be a Delzant polytope in $(\mathbb{R}^n)^{*}$[5] and with d facets.[6] Let $v_i \in \mathbb{Z}^n$, $i = 1, \ldots, d$, be the primitive[7] outward-pointing normal vectors to the facets of Δ. Then we can describe Δ as an intersection of halfspaces

$$\Delta = \{x \in (\mathbb{R}^n)^* \mid \langle x, v_i \rangle \leq \lambda_i,\ i = 1, \ldots, d\} \quad \text{for some } \lambda_i \in \mathbb{R}\ .$$

Example. For the picture on the next page, we have

$$\begin{aligned} \Delta &= \{x \in (\mathbb{R}^2)^* \mid x_1 \geq 0,\ x_2 \geq 0,\ x_1 + x_2 \leq 1\} \\ &= \{x \in (\mathbb{R}^2)^* \mid \langle x, (-1, 0) \rangle \leq 0\ ,\ \langle x, (0, -1) \rangle \leq 0\ ,\ \langle x, (1, 1) \rangle \leq 1\}\ . \end{aligned}$$

                                                                                        ◇

---

[5] Although we identify $\mathbb{R}^n$ with its dual via the euclidean inner product, it may be more clear to see Δ in $(\mathbb{R}^n)^*$ for Delzant's construction.

[6] A **face** of a polytope Δ is a set of the form $F = P \cap \{x \in \mathbb{R}^n \mid f(x) = c\}$ where $c \in \mathbb{R}$ and $f \in (\mathbb{R}^n)^*$ satisfies $f(x) \geq c, \forall x \in P$. A **facet** of an n-dimensional polytope is an $(n-1)$-dimensional face.

[7] A lattice vector $v \in \mathbb{Z}^n$ is **primitive** if it cannot be written as $v = ku$ with $u \in \mathbb{Z}^n$, $k \in \mathbb{Z}$ and $|k| > 1$; for instance, $(1, 1)$, $(4, 3)$, $(1, 0)$ are primitive, but $(2, 2)$, $(4, 6)$ are not.

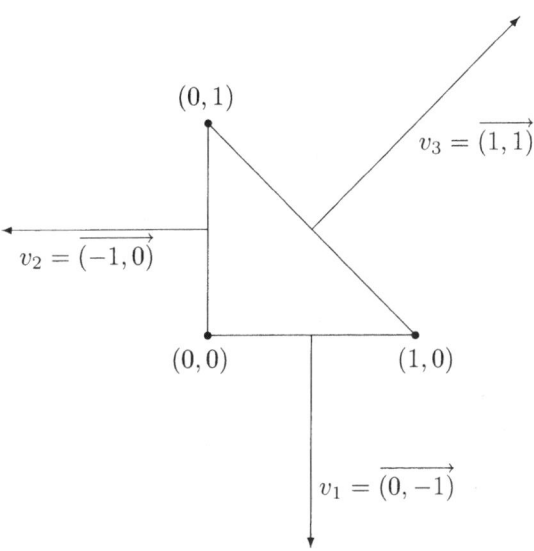

I.2.2 Orbit Spaces

Let $\psi : G \to \mathrm{Diff}(M)$ be any action.

Definition I.2.3. *The* **orbit** *of G through $p \in M$ is $\{\psi_g(p) \mid g \in G\}$. The* **stabilizer** *(or* isotropy*) of $p \in M$ is $G_p := \{g \in G \mid \psi_g(p) = p\}$.*

> **Exercise 13**
> If q is in the orbit of p, then G_q and G_p are conjugate subgroups.

Definition I.2.4. *We say that the action of G on M is:*

- **transitive** *if there is just one orbit,*

- **free** *if all stabilizers are trivial $\{e\}$,*

- **locally free** *if all stabilizers are discrete.*

Let $\sim$ be the orbit equivalence relation; for $p, q \in M$,

$$p \sim q \quad \Longleftrightarrow \quad p \text{ and } q \text{ are on the same orbit.}$$

The space of orbits $M/G := M/\sim$ is called the **orbit space**. Let

$$
\begin{array}{rcl}
\pi: & M & \longrightarrow & M/G \\
 & p & \longmapsto & \text{orbit through } p
\end{array}
$$

be the **point-orbit projection**.

We equip M/G with the weakest topology for which π is continuous, i.e., $\mathcal{U} \subseteq M/G$ is open if and only if $\pi^{-1}(\mathcal{U})$ is open in M. This is called the **quotient topology**. This topology can be "bad." For instance:

Example. Let $G = \mathbb{C}\backslash\{0\}$ act on $M = \mathbb{C}^n$ by

$$
\lambda \longmapsto \psi_\lambda = \text{multiplication by } \lambda .
$$

The orbits are the punctured complex lines (through non-zero vectors $z \in \mathbb{C}^n$), plus one "unstable" orbit through 0, which has a single point. The orbit space is

$$
M/G = \mathbb{P}^{n-1} \sqcup \{\text{point}\} .
$$

The quotient topology restricts to the usual topology on $\mathbb{P}^{n-1}$. The only open set containing $\{\text{point}\}$ in the quotient topology is the full space, hence the topology in M/G is *not* Hausdorff.

However, it suffices to remove 0 from $\mathbb{C}^n$ to obtain a Hausdorff orbit space: $\mathbb{P}^{n-1}$. Then there is also a compact (yet not complex) description of the orbit space by taking only unit vectors under the action of the circle subgroup:

$$
\mathbb{P}^{n-1} = \left(\mathbb{C}^n\backslash\{0\}\right)\Big/\left(\mathbb{C}\backslash\{0\}\right) = S^{2n-1}/S^1 .
$$

$\diamondsuit$

I.2.3   Symplectic Reduction

Let $\omega = \frac{i}{2}\sum dz_k \wedge d\bar{z}_k = \sum dx_k \wedge dy_k = \sum r_k dr_k \wedge d\theta_k$ be the standard symplectic form on $\mathbb{C}^n$. Consider the following S^1-action on $(\mathbb{C}^n, \omega)$:

$$
t \in S^1 \longmapsto \psi_t = \text{multiplication by } t .
$$

The action ψ is hamiltonian with moment map

$$
\begin{array}{rcl}
\mu: & \mathbb{C}^n & \longrightarrow & \mathbb{R} \\
 & z & \longmapsto & -\dfrac{\|z\|^2}{2} + \text{constant}
\end{array}
$$

since

$$
d\mu = -\tfrac{1}{2}d(\textstyle\sum r_k^2) ,
$$

$$
X^{\#} = \frac{\partial}{\partial\theta_1} + \frac{\partial}{\partial\theta_2} + \cdots + \frac{\partial}{\partial\theta_n} \quad \text{and} \quad
$$

$$
\imath_{X^{\#}}\omega = -\sum r_k dr_k = -\tfrac{1}{2}\sum dr_k^2 .
$$

If we choose the constant to be $\frac{1}{2}$, then $\mu^{-1}(0) = S^{2n-1}$ is the unit sphere. The orbit space of the zero level of the moment map is

$$\mu^{-1}(0)/S^1 = S^{2n-1}/S^1 = \mathbb{P}^{n-1} \ .$$

Moreover, this construction induces a symplectic form on $\mathbb{P}^{n-1}$, as a particular instance of the following major theorem.

Theorem I.2.5. (Marsden-Weinstein [34], Meyer [37]) *Let (M, ω, G, μ) be a hamiltonian G-space for a compact Lie group G. Let $i : \mu^{-1}(0) \hookrightarrow M$ be the inclusion map. Assume that G acts freely on $\mu^{-1}(0)$. Then*

(a) *the orbit space $M_{\mathrm{red}} = \mu^{-1}(0)/G$ is a manifold,*

(b) *$\pi : \mu^{-1}(0) \to M_{\mathrm{red}}$ is a principal G-bundle, and*

(c) *there is a symplectic form ω_{red} on M_{red} satisfying $i^*\omega = \pi^*\omega_{\mathrm{red}}$.*

For a proof of Theorem I.2.5, see for instance [12].

Definition I.2.6. *The pair $(M_{\mathrm{red}}, \omega_{\mathrm{red}})$ is called the **symplectic reduction** of (M, ω) with respect to G and μ (or the reduced space, or the symplectic quotient, or the Marsden-Weinstein-Meyer quotient, etc.).*

Remark. When M is Kähler, i.e., has a compatible complex structure, and the action of G preserves the complex structure, then the symplectic reduction has a natural Kähler structure. $\diamond$

Example. Consider the S^1-action on $(\mathbb{R}^{2n+2}, \omega_0)$ which, under the usual identification of $\mathbb{R}^{2n+2}$ with $\mathbb{C}^{n+1}$, corresponds to multiplication by e^{it}. This action is hamiltonian with a moment map $\mu : \mathbb{C}^{n+1} \to \mathbb{R}$ given by

$$\mu(z) = -\tfrac{1}{2}||z||^2 + \tfrac{1}{2} \ .$$

Symplectic reduction yields complex projective space $\mu^{-1}(0)/S^1 = \mathbb{P}^n$ equipped with the so-called **Fubini-Study symplectic form** $\omega_{\mathrm{red}} = \omega_{\mathrm{FS}}$. $\diamond$

Exercise 14
Recall that $\mathbb{P}^1 \simeq S^2$ as real 2-dimensional manifolds. Check that

$$\omega_{\mathrm{FS}} = \frac{1}{4}\omega_{\mathrm{standard}} \ ,$$

where $\omega_{\mathrm{standard}} = d\theta \wedge dh$ is the standard area form on S^2.

I.2.4   Extensions of Symplectic Reduction

We consider three basic extensions of the procedure of symplectic reduction.

1. **Reduction for product groups.**

   Let G_1 and G_2 be compact connected Lie groups and let $G = G_1 \times G_2$. Then $\mathfrak{g} \simeq \mathfrak{g}_1 \oplus \mathfrak{g}_2$ and $\mathfrak{g}^* \simeq \mathfrak{g}_1^* \oplus \mathfrak{g}_2^*$. Suppose that (M, ω, G, ν) is a hamiltonian G-space with moment map

   $$\nu : M \longrightarrow \mathfrak{g}_1^* \oplus \mathfrak{g}_2^* \ .$$

   Write $\nu = (\nu_1, \nu_2)$ where $\nu_i : M \to \mathfrak{g}_i^*$ for $i = 1, 2$. The fact that ν is equivariant implies that ν_1 is invariant under G_2 and ν_2 is invariant under G_1. Now reduce (M, ω) with respect to the G_1-action. Let

   $$Z_1 = \nu_1^{-1}(0) \ .$$

   Assume that G_1 acts freely on Z_1. Let $M_1 = Z_1 / G_1$ be the reduced space and let ω_1 be the corresponding reduced symplectic form. The action of G_2 on Z_1 commutes with the G_1-action. Since G_2 preserves ω, it follows that G_2 acts symplectically on (M_1, ω_1). Since G_1 preserves ν_2, G_1 also preserves $\nu_2 \circ \iota_1 : Z_1 \to \mathfrak{g}_2^*$, where $\iota_1 : Z_1 \hookrightarrow M$ is inclusion. Thus $\nu_2 \circ \iota_1$ is constant on fibers of $Z_1 \xrightarrow{p_1} M_1$. We conclude that there exists a smooth map $\mu_2 : M_1 \to \mathfrak{g}_2^*$ such that $\mu_2 \circ p_1 = \nu_2 \circ \iota_1$.

   > **Exercise 15**
   > Show that:
   >
   > (a) the map μ_2 is a moment map for the action of G_2 on (M_1, ω_1), and
   >
   > (b) if G acts freely on $\nu^{-1}(0,0)$, then G_2 acts freely on $\mu_2^{-1}(0)$, and there is a natural symplectomorphism
   >
   > $$\nu^{-1}(0,0)/G \ \simeq \ \mu_2^{-1}(0)/G_2 \ .$$

   This technique of performing reduction with respect to one factor of a product group at a time is called **reduction in stages**. It may be extended to reduction by a normal subgroup $H \subset G$ and by the corresponding quotient group G/H.

2. **Reduction at other levels.**

   Suppose that a compact Lie group G acts on a symplectic manifold (M, ω) in a hamiltonian way with moment map $\mu : M \to \mathfrak{g}^*$. Let $\xi \in \mathfrak{g}^*$. To reduce at the level ξ of μ, we need $\mu^{-1}(\xi)$ to be preserved by G, or else take the G-orbit of $\mu^{-1}(\xi)$, or equivalently take the inverse image $\mu^{-1}(\mathcal{O}_\xi)$ of the coadjoint orbit through ξ, or else take the quotient by the maximal subgroup of G which preserves $\mu^{-1}(\xi)$. Of course the level 0 is always preserved. Also, when G is a torus, any level is preserved and reduction at ξ for the moment map μ, is equivalent to reduction at 0 for a shifted moment map $\phi : M \to \mathfrak{g}^*$, $\phi(p) := \mu(p) - \xi$.

3. **Orbifold singularities.**

Roughly speaking, orbifolds (introduced by Satake in [42]) are singular mani-folds where each singularity is locally modeled on $\mathbb{R}^m/\Gamma$, for some finite group $\Gamma \subset \mathrm{GL}(m; \mathbb{R})$. For the precise definition, let $|M|$ be a Hausdorff topological space satisfying the second axiom of countability.

Definition I.2.7. *An **orbifold chart** on $|M|$ is a triple $(\mathcal{V}, \Gamma, \varphi)$, where $\mathcal{V}$ is a connected open subset of some euclidean space $\mathbb{R}^m$, Γ is a finite group which acts linearly on $\mathcal{V}$ so that the set of points where the action is not free has codimension at least two, and $\varphi : \mathcal{V} \to |M|$ is a Γ-invariant map inducing a homeomorphism from $\mathcal{V}/\Gamma$ onto its image $\mathcal{U} \subset |M|$. An **orbifold atlas** $\mathcal{A}$ for $|M|$ is a collection of orbifold charts on $|M|$ such that: the collection of images $\mathcal{U}$ forms a basis of open sets in $|M|$, and the charts are compatible in the sense that, whenever two charts $(\mathcal{V}_1, \Gamma_1, \varphi_1)$ and $(\mathcal{V}_2, \Gamma_2, \varphi_2)$ satisfy $\mathcal{U}_1 \subseteq \mathcal{U}_2$, there exists an injective homomorphism $\lambda : \Gamma_1 \to \Gamma_2$ and a λ-equivariant open embedding $\psi : \mathcal{V}_1 \to \mathcal{V}_2$ such that $\varphi_2 \circ \psi = \varphi_1$. Two orbifold atlases are **equivalent** if their union is still an atlas. An m-dimensional **orbifold** M is a Hausdorff topological space $|M|$ satisfying the second axiom of countability, plus an equivalence class of orbifold atlases on $|M|$.*

Notice that we do not require the action of each group Γ to be effective. Given a point p on an orbifold M, let $(\mathcal{V}, \Gamma, \varphi)$ be an orbifold chart for a neighborhood $\mathcal{U}$ of p. The **orbifold structure group** of p, Γ_p, is (the isomor-phism class of) the isotropy group of a pre-image of p under ϕ. We may always choose an orbifold chart $(\mathcal{V}, \Gamma, \varphi)$ such that $\varphi^{-1}(p)$ is a single point (which is fixed by Γ). In this case $\Gamma \simeq \Gamma_p$, and we say that $(\mathcal{V}, \Gamma, \varphi)$ is a **structure chart** for p.

An ordinary manifold is a special case of orbifold where each group Γ is the identity group. Quotients of manifolds by locally free actions of Lie groups are orbifolds. In fact, any orbifold M has a presentation of this form obtained as follows. Given a structure chart $(\mathcal{V}, \Gamma, \varphi)$ for $p \in M$ with image $\mathcal{U}$, the **orbifold tangent space** at p is the quotient of the tangent space to $\mathcal{V}$ at $\varphi^{-1}(p)$ by the induced action of Γ:

$$T_p M := T_{\varphi^{-1}(p)} \mathcal{V}/\Gamma \ .$$

The collection of the orbifold tangent spaces at all p, builds up the **orbifold tangent bundle** TM, which has a natural structure of smooth manifold out-side the zero section. The general linear group $\mathrm{GL}(m; \mathbb{R})$ acts locally freely on $TM \setminus 0$, and $M \simeq (TM - 0)/\mathrm{GL}(m; \mathbb{R})$. Choosing a riemannian metric and taking the orthonormal frame bundle, $O(TM)$, we present M as $O(TM)/O(m)$.

Example. Let $G = \mathbb{T}^n$ be an n-torus acting on a symplectic manifold (M, ω) in a hamiltonian way with moment map $\mu : M \rightarrow \mathfrak{g}^*$. For any $\xi \in \mathfrak{g}^*$, the level $\mu^{-1}(\xi)$ is preserved by the $\mathbb{T}^n$-action. Suppose that ξ is a regular value of μ.[8] Then $\mu^{-1}(\xi)$ is a submanifold of codimension n. Let G_p be the stabilizer of p, and $\mathfrak{g}_p$ its Lie algebra. Note that

$$
\begin{aligned}
\xi \text{ regular} \quad &\Longleftrightarrow \quad d\mu_p \text{ is surjective at all } p \in \mu^{-1}(\xi) \\
&\Longleftrightarrow \quad \mathfrak{g}_p = 0 \text{ for all } p \in \mu^{-1}(\xi) \\
&\Longleftrightarrow \quad \text{the stabilizers on } \mu^{-1}(\xi) \text{ are finite .}
\end{aligned}
$$

By the slice theorem (see, for instance, [6, 12]), near $\mathcal{O}_p$ the orbit space $\mu^{-1}(\xi)/G$ is modeled by S/G_p, where S is a G_p-invariant disk in $\mu^{-1}(\xi)$ through p and transverse to $\mathcal{O}_p$. Hence, $\mu^{-1}(\xi)/G$ is an orbifold. $\qquad \diamond$

Example. Consider the S^1-action on $\mathbb{C}^2$ by $e^{i\theta} \cdot (z_1, z_2) = (e^{ik\theta} z_1, e^{i\theta} z_2)$ for some fixed integer $k \geq 2$. This is hamiltonian with moment map

$$
\begin{array}{cccc}
\mu : & \mathbb{C}^2 & \longrightarrow & \mathbb{R} \\
 & (z_1, z_2) & \longmapsto & -\frac{1}{2}(k|z_1|^2 + |z_2|^2) .
\end{array}
$$

Any $\xi < 0$ is a regular value and $\mu^{-1}(\xi)$ is a 3-dimensional ellipsoid. The stabilizer of $(z_1, z_2) \in \mu^{-1}(\xi)$ is $\{1\}$ if $z_2 \neq 0$, and is

$$
\mathbb{Z}_k = \left\{ e^{i\frac{2\pi\ell}{k}} \mid \ell = 0, 1, \ldots, k-1 \right\}
$$

if $z_2 = 0$. The reduced space $\mu^{-1}(\xi)/S^1$ is called a **teardrop** orbifold or *conehead*; it has one **cone** (also known as a *dunce cap*) singularity with cone angle $\frac{2\pi}{k}$, that is, a point with orbifold structure group $\mathbb{Z}_k$. $\qquad \diamond$

Example. Let S^1 act on $\mathbb{C}^2$ by $e^{i\theta} \cdot (z_1, z_2) = (e^{ik\theta} z_1, e^{i\ell\theta} z_2)$ for some integers $k, \ell \geq 2$. Suppose that k and ℓ are relatively prime. Then

$$
\begin{array}{lll}
(z_1, 0) & \text{has stabilizer } \mathbb{Z}_k & (\text{for } z_1 \neq 0) , \\
(0, z_2) & \text{has stabilizer } \mathbb{Z}_\ell & (\text{for } z_2 \neq 0) , \\
(z_1, z_2) & \text{has stabilizer } \{1\} & (\text{for } z_1, z_2 \neq 0) .
\end{array}
$$

The quotient $\mu^{-1}(\xi)/S^1$ is called a **football** orbifold. It has two cone singularities, one with angle $\frac{2\pi}{k}$ and another with angle $\frac{2\pi}{\ell}$. $\qquad \diamond$

---

[8] By Sard's theorem, the singular values of μ form a set of measure zero.

Example. More generally, the reduced spaces of S^1 acting on $\mathbb{C}^n$ by

$$e^{i\theta} \cdot (z_1, \ldots, z_n) = (e^{ik_1\theta} z_1, \ldots, e^{ik_n\theta} z_n) \ ,$$

are called **weighted** (or *twisted*) **projective spaces**. $\diamond$

The differential-geometric notions of vector fields, differential forms, exterior differentiation, group actions, etc., extend naturally to orbifolds by gluing corresponding local Γ-invariant or Γ-equivariant objects. In particular, a **symplectic orbifold** is a pair (M, ω) where M is an orbifold and ω is a closed 2-form on M which is nondegenerate at every point of M.

Definition I.2.8. *A **symplectic toric orbifold** is a compact connected symplectic orbifold (M, ω) equipped with an effective hamiltonian action of a torus $\mathbb{T}$ of dimension equal to half the dimension of the orbifold,*

$$\dim \mathbb{T} = \frac{1}{2} \dim M \ ,$$

and with a choice of a corresponding moment map μ.

Symplectic toric orbifolds have been classified by Lerman and Tolman [33] in a theorem which generalizes Delzant's theorem: a symplectic toric orbifold is determined by its moment polytope plus a positive integer label attached to each of the polytope facets. The polytopes which occur in the Lerman-Tolman classification are more general than the Delzant polytopes in the sense that only simplicity and rationality are required; the edge vectors $u_1, \ldots, u_n$ need only form a rational basis of $\mathbb{Z}^n$. In the case where the integer labels are all equal to 1, the failure of the polytope smoothness accounts for all orbifold singularities. Throughout the rest of these notes, we concentrate on the manifold case.

I.2.5 Delzant's Construction

Following [15, 24], we prove the existence part (or surjectivity) in Delzant's theorem, by using symplectic reduction to associate to an n-dimensional Delzant polytope Δ a symplectic toric manifold $(M_\Delta, \omega_\Delta, \mathbb{T}^n, \mu_\Delta)$.

Let Δ be a Delzant polytope with d facets. Let $v_i \in \mathbb{Z}^n$, $i = 1, \ldots, d$, be the primitive outward-pointing normal vectors to the facets. For some $\lambda_i \in \mathbb{R}$, we can write

$$\Delta = \{x \in (\mathbb{R}^n)^* \mid \langle x, v_i \rangle \leq \lambda_i, \ i = 1, \ldots, d\} \ .$$

Let $e_1 = (1, 0, \ldots, 0), \ldots, e_d = (0, \ldots, 0, 1)$ be the standard basis of $\mathbb{R}^d$. Consider

$$\begin{array}{rccc} \pi : & \mathbb{R}^d & \longrightarrow & \mathbb{R}^n \\ & e_i & \longmapsto & v_i \ . \end{array}$$

Lemma I.2.9. *The map π is onto and maps $\mathbb{Z}^d$ onto $\mathbb{Z}^n$.*

Proof. The set $\{e_1, \ldots, e_d\}$ is a basis of $\mathbb{Z}^d$. The set $\{v_1, \ldots, v_d\}$ spans $\mathbb{Z}^n$ for the following reason. At a vertex p, the edge vectors $u_1, \ldots, u_n \in (\mathbb{R}^n)^*$, form a basis for $(\mathbb{Z}^n)^*$ which, by a change of basis if necessary, we may assume is the standard basis. Then the corresponding primitive normal vectors to the facets meeting at p are symmetric (in the sense of multiplication by -1) to the u_i's, hence form a basis of $\mathbb{Z}^n$.   □

Therefore, π induces a surjective map, still called π, between tori:

$$
\begin{array}{ccc}
\mathbb{R}^d/(2\pi\mathbb{Z}^d) & \xrightarrow{\ \pi\ } & \mathbb{R}^n/(2\pi\mathbb{Z}^n) \\
\| & & \| \\
\mathbb{T}^d & \longrightarrow & \mathbb{T}^n \qquad \longrightarrow \quad 1 \ .
\end{array}
$$

The kernel N of π is a $(d-n)$-dimensional Lie subgroup of $\mathbb{T}^d$ with inclusion $i : N \hookrightarrow \mathbb{T}^d$. Let $\mathfrak{n}$ be the Lie algebra of N. The exact sequence of tori

$$
1 \longrightarrow N \xrightarrow{\ i\ } \mathbb{T}^d \xrightarrow{\ \pi\ } \mathbb{T}^n \longrightarrow 1
$$

induces an exact sequence of Lie algebras

$$
0 \longrightarrow \mathfrak{n} \xrightarrow{\ i\ } \mathbb{R}^d \xrightarrow{\ \pi\ } \mathbb{R}^n \longrightarrow 0
$$

with dual exact sequence

$$
0 \longrightarrow (\mathbb{R}^n)^* \xrightarrow{\ \pi^*\ } (\mathbb{R}^d)^* \xrightarrow{\ i^*\ } \mathfrak{n}^* \longrightarrow 0 \ .
$$

Now consider $\mathbb{C}^d$ with symplectic form $\omega_0 = \frac{i}{2} \sum dz_k \wedge d\bar{z}_k$, and standard hamiltonian action of $\mathbb{T}^d$ given by

$$
(e^{it_1}, \ldots, e^{it_d}) \cdot (z_1, \ldots, z_d) = (e^{it_1} z_1, \ldots, e^{it_d} z_d) \ .
$$

The moment map is $\phi : \mathbb{C}^d \longrightarrow (\mathbb{R}^d)^*$ defined by

$$
\phi(z_1, \ldots, z_d) = -\frac{1}{2}(|z_1|^2, \ldots, |z_d|^2) + \text{constant} \ ,
$$

where we choose the constant to be $(\lambda_1, \ldots, \lambda_d)$. By Exercise 6, the subtorus N acts on $\mathbb{C}^d$ in a hamiltonian way with moment map

$$
i^* \circ \phi : \mathbb{C}^d \longrightarrow \mathfrak{n}^* \ .
$$

Let $Z = (i^* \circ \phi)^{-1}(0)$ be the zero-level set.

Claim 1. The set Z is compact and N acts freely on Z.

We postpone the proof of this claim until further down.

Since i^* is surjective, $0 \in \mathfrak{n}^*$ is a regular value of $i^* \circ \phi$. Hence, Z is a compact submanifold of $\mathbb{C}^d$ of (real) dimension $2d - (d - n) = d + n$. The orbit space $M_\Delta = Z/N$ is a compact manifold of (real) dimension $\dim Z - \dim N = (d + n) - (d - n) = 2n$. The point-orbit map $p : Z \to M_\Delta$ is a principal N-bundle over M_Δ. Consider the diagram

$$
\begin{array}{ccc}
Z & \overset{j}{\hookrightarrow} & \mathbb{C}^d \\
p \downarrow & & \\
M_\Delta & &
\end{array}
$$

where $j : Z \hookrightarrow \mathbb{C}^d$ is inclusion. The Marsden-Weinstein-Meyer theorem guarantees the existence of a symplectic form ω_Δ on M_Δ satisfying

$$
p^* \omega_\Delta = j^* \omega_0 .
$$

Since Z is connected, the compact symplectic $2n$-dimensional manifold $(M_\Delta, \omega_\Delta)$ is also connected.

Proof of Claim 1. The set Z is clearly closed, hence in order to show that it is compact it suffices (by the Heine-Borel theorem) to show that Z is bounded. Let Δ' be the image of Δ by π^*. We will show that $\phi(Z) = \Delta'$.

Lemma I.2.10. *Let* $y \in (\mathbb{R}^d)^*$*. Then:*

$$
y \in \Delta' \iff y \text{ is in the image of } Z \text{ by } \phi .
$$

Proof. The value y is in the image of Z by ϕ if and only if both of the following conditions hold:

1. y is in the image of ϕ;

2. $i^* y = 0$.

Using the expression for ϕ and the third exact sequence, we see that these conditions are equivalent to:

1. $\langle y, e_i \rangle \leq \lambda_i$ for $i = 1, \ldots, d$;

2. $y = \pi^*(x)$ for some $x \in (\mathbb{R}^n)^*$.

Suppose that the second condition holds, so that $y = \pi^*(x)$. Then

$$
\begin{array}{rcl}
\langle y, e_i \rangle \leq \lambda_i, \forall i & \iff & \langle \pi^*(x), e_i \rangle \leq \lambda_i, \forall i \\
& \iff & \langle x, \pi(e_i) \rangle \leq \lambda_i, \forall i \\
& \iff & \langle x, v_i \rangle \leq \lambda_i, \forall i \\
& \iff & x \in \Delta .
\end{array}
$$

Thus, $y \in \phi(Z) \iff y \in \pi^*(\Delta) = \Delta'$. $\square$

Since we have that Δ' is compact, that ϕ is a proper map and that $\phi(Z) = \Delta'$, we conclude that Z must be bounded, and hence compact.

It remains to show that N acts freely on Z.

Pick a vertex p of Δ, and let $I = \{i_1, \ldots, i_n\}$ be the set of indices for the n facets meeting at p. Pick $z \in Z$ such that $\phi(z) = \pi^*(p)$. Then p is characterized by n equations $\langle p, v_i \rangle = \lambda_i$ where i ranges in I:

$$\langle p, v_i \rangle = \lambda_i \quad \iff \quad \langle p, \pi(e_i) \rangle = \lambda_i$$
$$\iff \quad \langle \pi^*(p), e_i \rangle = \lambda_i$$
$$\iff \quad \langle \phi(z), e_i \rangle = \lambda_i$$
$$\iff \quad i\text{-th coordinate of } \phi(z) \text{ is equal to } \lambda_i$$
$$\iff \quad -\tfrac{1}{2}|z_i|^2 + \lambda_i = \lambda_i$$
$$\iff \quad z_i = 0 .$$

Hence, those z's are points whose coordinates in the set I are zero, and whose other coordinates are nonzero. Without loss of generality, we may assume that $I = \{1, \ldots, n\}$. The stabilizer of z is

$$(\mathbb{T}^d)_z = \{(t_1, \ldots, t_n, 1, \ldots, 1) \in \mathbb{T}^d\} .$$

As the restriction $\pi : (\mathbb{R}^d)_z \to \mathbb{R}^n$ maps the vectors $e_1, \ldots, e_n$ to a $\mathbb{Z}$-basis $v_1, \ldots, v_n$ of $\mathbb{Z}^n$ (respectively), at the level of groups, $\pi : (\mathbb{T}^d)_z \to \mathbb{T}^n$ must be bijective. Since $N = \ker(\pi : \mathbb{T}^d \to \mathbb{T}^n)$, we conclude that $N \cap (\mathbb{T}^d)_z = \{e\}$, i.e., $N_z = \{e\}$. Hence all N-stabilizers at points mapping to vertices are trivial. But this was the worst case, since other stabilizers $N_{z'}$ ($z' \in Z$) are contained in stabilizers for points z which map to vertices. This concludes the proof of Claim 1.
$\square$

Given a Delzant polytope Δ, we have constructed a symplectic manifold $(M_\Delta, \omega_\Delta)$ where $M_\Delta = Z/N$ is a compact $2n$-dimensional manifold and ω_Δ is the reduced symplectic form.

Claim 2. The manifold $(M_\Delta, \omega_\Delta)$ is a hamiltonian $\mathbb{T}^n$-space with a moment map μ_Δ having image $\mu_\Delta(M_\Delta) = \Delta$.

Proof of Claim 2. Let z be such that $\phi(z) = \pi^*(p)$ where p is a vertex of Δ, as in the proof of Claim 1. Let $\sigma : \mathbb{T}^n \to (\mathbb{T}^d)_z$ be the inverse for the earlier bijection $\pi : (\mathbb{T}^d)_z \to \mathbb{T}^n$. Since we have found a *section*, i.e., a right inverse for π, in the exact sequence

$$1 \longrightarrow N \overset{i}{\longrightarrow} \mathbb{T}^d \underset{\underset{\sigma}{\longleftarrow}}{\overset{\pi}{\longrightarrow}} \mathbb{T}^n \longrightarrow 1 ,$$

the exact sequence *splits*, i.e., becomes like a sequence for a product, as we obtain an isomorphism

$$(i, \sigma) : N \times \mathbb{T}^n \overset{\simeq}{\longrightarrow} \mathbb{T}^d .$$

The action of the $\mathbb{T}^n$ factor (or, more rigorously, $\sigma(\mathbb{T}^n) \subset \mathbb{T}^d$) descends to the quotient $M_\Delta = Z/N$.

It remains to show that the $\mathbb{T}^n$-action on M_Δ is hamiltonian with appropriate moment map.

Consider the diagram

$$Z \xrightarrow{j} \mathbb{C}^d \xrightarrow{\phi} (\mathbb{R}^d)^* \simeq \eta^* \oplus (\mathbb{R}^n)^* \xrightarrow{\sigma^*} (\mathbb{R}^n)^*$$
$$p \downarrow$$
$$M_\Delta$$

where the last horizontal map is simply projection onto the second factor. Since the composition of the horizontal maps is constant along N-orbits, it descends to a map

$$\mu_\Delta : M_\Delta \longrightarrow (\mathbb{R}^n)^*$$

which satisfies

$$\mu_\Delta \circ p = \sigma^* \circ \phi \circ j \ .$$

By Exercise 15 on reduction for product groups, this is a moment map for the action of $\mathbb{T}^n$ on $(M_\Delta, \omega_\Delta)$. Finally, the image of μ_Δ is:

$$\mu_\Delta(M_\Delta) = (\mu_\Delta \circ p)(Z) = (\sigma^* \circ \phi \circ j)(Z) = (\sigma^* \circ \pi^*)(\Delta) = \Delta \ ,$$

because $\phi(Z) = \pi^*(\Delta)$ and $\sigma^* \circ \pi^* = (\pi \circ \sigma)^* = \mathrm{id}$.

We conclude that $(M_\Delta, \omega_\Delta, \mathbb{T}^n, \mu_\Delta)$ is the required toric manifold corresponding to Δ. $\qquad\qquad\square$

Exercise 16

Let Δ be an n-dimensional Delzant polytope, and let $(M_\Delta, \omega_\Delta, \mathbb{T}'', \mu_\Delta)$ be the associated symplectic toric manifold. Show that μ_Δ maps the fixed points of $\mathbb{T}^n$ bijectively onto the vertices of Δ.

By the remark in Section I.2.3, Delzant's construction yields a natural Kähler structure on each symplectic toric manifold.

I.2.6 Idea Behind Delzant's Construction

The space $\mathbb{R}^d$ is *universal* in the sense that any n-dimensional polytope Δ with d facets can be obtained by intersecting the negative orthant $\mathbb{R}^d_-$ with an affine plane A. Given Δ, to construct A first write Δ as:

$$\Delta = \{x \in \mathbb{R}^n \mid \langle x, v_i \rangle \le \lambda_i, \ i = 1, \ldots, d\} \ .$$

Define

$$\pi : \quad \mathbb{R}^d \quad \longrightarrow \quad \mathbb{R}^n \qquad \text{with dual map} \qquad \pi^* : \quad \mathbb{R}^n \quad \longrightarrow \quad \mathbb{R}^d .$$
$$e_i \quad \longmapsto \quad v_i$$

Then $\pi^* - \lambda : \mathbb{R}^n \longrightarrow \mathbb{R}^d$ is an affine map, where $\lambda = (\lambda_1, \ldots, \lambda_d)$. Let A be the image of $\pi^* - \lambda$. Then A is an n-dimensional affine space.

Lemma I.2.11. *We have the equality* $(\pi^* - \lambda)(\Delta) = \mathbb{R}^d_- \cap A.$

Proof. Let $x \in \mathbb{R}^n$. Then

$$
\begin{aligned}
(\pi^* - \lambda)(x) \in \mathbb{R}^d_- \quad &\Longleftrightarrow \quad \langle \pi^*(x) - \lambda, e_i \rangle \leq 0, \forall i \\
&\Longleftrightarrow \quad \langle x, \pi(e_i) \rangle - \lambda_i \leq 0, \forall i \\
&\Longleftrightarrow \quad \langle x, v_i \rangle \leq \lambda_i, \forall i \\
&\Longleftrightarrow \quad x \in \Delta .
\end{aligned}
$$

$\square$

We conclude that $\Delta \simeq \mathbb{R}^d_- \cap A$. Now $\mathbb{R}^d_-$ is the image of the moment map for the standard hamiltonian action of $\mathbb{T}^d$ on $\mathbb{C}^d$

$$\phi : \mathbb{C}^d \quad \longrightarrow \quad \mathbb{R}^d$$
$$(z_1, \ldots, z_d) \quad \longmapsto \quad -\tfrac{1}{2}(|z_1|^2, \ldots, |z_d|^2) .$$

Facts.

- The set $\phi^{-1}(A) \subset \mathbb{C}^d$ is a compact submanifold. Let $i : \phi^{-1}(A) \hookrightarrow \mathbb{C}^d$ denote inclusion. Then $i^* \omega_0$ is a closed 2-form which is degenerate. Its kernel is an integrable distribution. The corresponding foliation is called the **null foliation**.

- The null foliation of $i^* \omega_0$ is a principal fibration, so we take the quotient:

$$N \quad \hookrightarrow \quad \phi^{-1}(A)$$
$$\downarrow$$
$$M_\Delta \quad := \quad \phi^{-1}(A)/N .$$

Let ω_Δ be the reduced symplectic form.

- The (non-effective) action of $\mathbb{T}^d = N \times \mathbb{T}^n$ on $\phi^{-1}(A)$ has a "moment map" with image $\phi(\phi^{-1}(A)) = \Delta$. (By "moment map" we mean a map satisfying the usual definition even though the closed 2-form is not symplectic.)

There is a remaining action of $\mathbb{T}^n$ on M_Δ which is hamiltonian with a moment map $\mu_\Delta : M_\Delta \to \mathbb{R}^n$ defined by the commutative diagram

$$
\begin{array}{ccccc}
\phi^{-1}(A) & \overset{j}{\hookrightarrow} & \mathbb{C}^d & \overset{\phi}{\longrightarrow} & \mathbb{R}^d \\
p \downarrow & & & & \downarrow \mathrm{pr}_2 \\
M_\Delta & & \overset{\mu_\Delta}{\dashrightarrow} & & \mathbb{R}^n
\end{array}
$$

where $\mathrm{pr}_2 : \mathbb{T}^d = N \times \mathbb{T}^n \to \mathbb{T}^n$ is projection onto the second factor.

Theorem I.2.12. *For any $x \in \Delta$, we have that $\mu_\Delta^{-1}(x)$ is a single $\mathbb{T}^n$-orbit.*

Proof. Exercise: First consider the standard $\mathbb{T}^d$-action on $\mathbb{C}^d$ with moment map $\phi : \mathbb{C}^d \to \mathbb{R}^d$. Show that $\phi^{-1}(y)$ is a single $\mathbb{T}^d$-orbit for any $y \in \phi(\mathbb{C}^d)$. Now observe that

$$
y \in \Delta' = \pi^*(\Delta) \iff \phi^{-1}(y) \subseteq Z .
$$

Suppose that $y = \pi^*(x)$. Show that $\mu_\Delta^{-1}(x) = \phi^{-1}(y)/N$. But $\phi^{-1}(y)$ is a single $\mathbb{T}^d$-orbit where $\mathbb{T}^d = N \times \mathbb{T}^n$, hence $\mu_\Delta^{-1}(x)$ is a single $\mathbb{T}^n$-orbit. $\square$

Therefore, for toric manifolds, Δ is the orbit space.

Now Δ is a *manifold with corners*. At every point p in a face F, the tangent space $T_p\Delta$ is the subspace of $\mathbb{R}^n$ tangent to F. We can visualize $(M_\Delta, \omega_\Delta, \mathbb{T}^n, \mu_\Delta)$ from Δ as follows. First take the product $\mathbb{T}^n \times \Delta$. Let p lie in the interior of $\mathbb{T}^n \times \Delta$. The tangent space at p is $\mathbb{R}^n \times (\mathbb{R}^n)^*$. Define ω_p by:

$$
\omega_p(v, \xi) = \xi(v) = -\omega_p(\xi, v) \quad \text{and} \quad \omega_p(v, v') = \omega(\xi, \xi') = 0 .
$$

for all $v, v' \in \mathbb{R}^n$ and $\xi, \xi' \in (\mathbb{R}^n)^*$. Then ω is a closed nondegenerate 2-form on the interior of $\mathbb{T}^n \times \Delta$. At the corner there are directions missing in $(\mathbb{R}^n)^*$, so ω is a degenerate pairing. Hence, we need to eliminate the corresponding directions in $\mathbb{R}^n$. To do this, we collapse the orbits corresponding to subgroups of $\mathbb{T}^n$ generated by directions orthogonal to the annihilator of that face.

Example. Consider

$$
(S^2, \omega = d\theta \wedge dh, S^1, \mu = h) ,
$$

where S^1 acts on S^2 by rotation. The image of μ is the line segment $I = [-1, 1]$. The product $S^1 \times I$ is an open-ended cylinder. By collapsing each end of the cylinder to a point, we recover the 2-sphere. $\diamond$

Exercise 17
Build $\mathbb{P}^2$ from $\mathbb{T}^2 \times \Delta$ where Δ is a right-angled isosceles triangle.

Finally, $\mathbb{T}^n$ acts on $\mathbb{T}^n \times \Delta$ by multiplication on the $\mathbb{T}^n$ factor. The moment map for this action is projection onto the Δ factor.

Exercise 18

Follow through the details of Delzant's construction for the case of $\Delta = [0, a] \subset \mathbb{R}^*$ ($n = 1, d = 2$). Let $v(= 1)$ be the standard basis vector in $\mathbb{R}$. Then Δ is described by

$$\langle x, -v \rangle \leq 0 \qquad \text{and} \qquad \langle x, v \rangle \leq a ,$$

where $v_1 = -v$, $v_2 = v$, $\lambda_1 = 0$ and $\lambda_2 = a$.

The projection

$$\begin{array}{ccc} \mathbb{R}^2 & \xrightarrow{\pi} & \mathbb{R} \\ e_1 & \longmapsto & -v \\ e_2 & \longmapsto & v \end{array}$$

has kernel equal to the span of $(e_1 + e_2)$, so that N is the diagonal subgroup of $\mathbb{T}^2 = S^1 \times S^1$. The exact sequences become

$$\begin{array}{ccccccccc} 1 & \longrightarrow & N & \xrightarrow{i} & \mathbb{T}^2 & \xrightarrow{\pi} & S^1 & \longrightarrow & 1 \\ & & t & \longmapsto & (t, t) & & & & \\ & & & & (t_1, t_2) & \longmapsto & t_1^{-1} t_2 & & \end{array}$$

$$\begin{array}{ccccccccc} 0 & \longrightarrow & \mathfrak{n} & \xrightarrow{i} & \mathbb{R}^2 & \xrightarrow{\pi} & \mathbb{R} & \longrightarrow & 0 \\ & & x & \longmapsto & (x, x) & & & & \\ & & & & (x_1, x_2) & \longmapsto & x_2 - x_1 & & \end{array}$$

$$\begin{array}{ccccccccc} 0 & \longrightarrow & \mathbb{R}^* & \xrightarrow{\pi^*} & (\mathbb{R}^2)^* & \xrightarrow{i^*} & \mathfrak{n}^* & \longrightarrow & 0 \\ & & x & \longmapsto & (-x, x) & & & & \\ & & & & (x_1, x_2) & \longmapsto & x_1 + x_2 . & & \end{array}$$

The action of the diagonal subgroup $N = \{(e^{it}, e^{it}) \in S^1 \times S^1\}$ on $\mathbb{C}^2$,

$$(e^{it}, e^{it}) \cdot (z_1, z_2) = (e^{it} z_1, e^{it} z_2) ,$$

has moment map

$$(i^* \circ \phi)(z_1, z_2) = -\tfrac{1}{2}(|z_1|^2 + |z_2|^2) + a ,$$

with zero-level set

$$(i^* \circ \phi)^{-1}(0) = \{(z_1, z_2) \in \mathbb{C}^2 : |z_1|^2 + |z_2|^2 = 2a\} .$$

Hence, the reduced space is a projective space:

$$(i^* \circ \phi)^{-1}(0)/N = \mathbb{P}^1 .$$

Exercise 19

Consider the standard $(S^1)^3$-action on $\mathbb{P}^3$:

$$(e^{i\theta_1}, e^{i\theta_2}, e^{i\theta_3}) \cdot [z_0 : z_1 : z_2 : z_3] = [z_0 : e^{i\theta_1} z_1 : e^{i\theta_2} z_2 : e^{i\theta_3} z_3] .$$

Exhibit explicitly the subsets of $\mathbb{P}^3$ for which the stabilizer under this action is $\{1\}$, S^1, $(S^1)^2$ and $(S^1)^3$. Show that the images of these subsets under the moment map are the interior, the facets, the edges and the vertices, respectively.

Exercise 20

What would be the classification of symplectic toric manifolds if, instead of the equivalence relation defined in Section I.1.6, one considered to be equivalent those $(M_i, \omega_i, \mathbb{T}_i, \mu_i)$, $i = 1, 2$, related by an isomorphism $\lambda : \mathbb{T}_1 \to \mathbb{T}_2$ and a λ-equivariant symplectomorphism $\varphi : M_1 \to M_2$ such that:

(a) the maps μ_1 and $\mu_2 \circ \varphi$ are equal up to a constant?

(b) we have $\mu_1 = \ell \circ \mu_2 \circ \varphi$ for some $\ell \in SL(n; \mathbb{Z})$?

Exercise 21

(a) Classify all 2-dimensional Delzant polytopes with 3 vertices, i.e., triangles, up to translation, change of scale and the action of $SL(2; \mathbb{Z})$.

Hint: By a linear transformation in $SL(2; \mathbb{Z})$, we can make one of the angles in the polytope into a square angle. How are the lengths of the two edges forming that angle related?

(b) Classify all 2-dimensional Delzant polytopes with 4 vertices, up to translation and the action of $SL(2; \mathbb{Z})$.

Hint: By a linear transformation in $SL(2; \mathbb{Z})$, we can make one of the angles in the polytope into a square angle. Check that automatically another angle also becomes 90°.

(c) What are all the 4-dimensional symplectic toric manifolds that have four fixed points?

Exercise 22

Let Δ be the n-simplex in $\mathbb{R}^n$ spanned by the origin and the standard basis vectors $(1, 0, \ldots, 0), \ldots, (0, \ldots, 0, 1)$. Show that the corresponding symplectic toric manifold is projective space, $M_\Delta = \mathbb{P}^n$.

Exercise 23

Which $2n$-dimensional toric manifolds have exactly $n + 1$ fixed points?

I.3   Moment Polytopes

The general theme behind this and Lecture 6 is *how to understand a toric manifold from its polytope.* After reviewing the basics of Morse theory following [39], we compute the homology of symplectic toric manifolds using Morse theory; an appropriate Morse function is provided by a moment map with respect to a suitable circle subgroup. We go on to describe elementary surgery constructions based on symplectic reduction, which hold in the category of symplectic toric manifolds.

I.3.1   Equivariant Darboux Theorem

The following two theorems describe standard neighborhoods of fixed points. Their proofs rely on the equivariant version of the Moser trick and may be found in [27].

Theorem I.3.1. (Equivariant Darboux)   *Let (M, ω) be a $2n$-dimensional symplectic manifold equipped with a symplectic action of a compact Lie group G, and let q be a fixed point. Then there exists a G-invariant chart $(\mathcal{U}, x_1, \ldots, x_n, y_1, \ldots, y_n)$ centered at q and G-equivariant with respect to a linear action of G on $\mathbb{R}^{2n}$ such that*

$$\omega|_{\mathcal{U}} = \sum_{k=1}^{n} dx_k \wedge dy_k \ .$$

A suitable linear action on $\mathbb{R}^{2n}$ is equivalent to the induced action of G on $T_q M$. In particular, if G is a torus, this linear action is characterized by the weights occuring in the representation of G on $T_q M$. Now any symplectic action is locally hamiltonian. In order to prepare the computation of the Betti numbers of a symplectic toric manifold by using a moment map as a Morse function, we next specify the local picture for a moment map near a fixed point of a hamiltonian torus action.

Theorem I.3.2.   *Let $(M^{2n}, \omega, \mathbb{T}^m, \mu)$ be a hamiltonian $\mathbb{T}^m$-space, where q is a fixed point. Then there exists a chart $(\mathcal{U}, x_1, \ldots, x_n, y_1, \ldots, y_n)$ centered at q and weights $\lambda^{(1)}, \ldots, \lambda^{(n)} \in \mathbb{Z}^m$ such that*

(a)

$$\omega|_{\mathcal{U}} = \sum_{k=1}^{n} dx_k \wedge dy_k \ , \qquad and$$

(b)

$$\mu|_{\mathcal{U}} = \mu(q) - \frac{1}{2} \sum_{k=1}^{n} \lambda^{(k)} (x_k^2 + y_k^2) \ .$$

This theorem guarantees the existence of a Darboux chart centered at any fixed point where the moment map looks like the moment map for a linear action on $\mathbb{R}^{2n}$. In other words, the real analogue of the model in Exercise 9 is a general local picture near a fixed point of a hamiltonian torus action.

> **Exercise 24**
> Show that for a symplectic toric manifold the weights $\lambda^{(1)}, \ldots, \lambda^{(n)}$ form a $\mathbb{Z}$-basis of $\mathbb{Z}^m$.

As a consequence of Theorem I.3.2 and of the previous exercise, the bijection claimed in Delzant's theorem is well-defined. Indeed, each vertex of a moment polytope satisfies the simplicity, rationality and smoothness conditions which characterize Delzant polytopes.

Another consequence of Theorem I.3.2 is that a moment map for a symplectic toric manifold yields a lot of *Morse functions*, as we will next explore.

I.3.2 Morse Theory

Let M be an m-dimensional manifold. A smooth function $f : M \to \mathbb{R}$ is a **Morse function** on M if all of its critical points are nondegenerate.[9]

The **index** of a bilinear function $H : \mathbb{R}^m \times \mathbb{R}^m \to \mathbb{R}$ is the maximal dimension of a subspace of $\mathbb{R}$ where H is negative definite. The **nullity** of H is the dimension of its nullspace, that is, the subspace consisting of all $v \in \mathbb{R}^m$ such that $H(v, w) = 0$ for all $w \in \mathbb{R}^m$. Hence, a critical point q of $f : M \to \mathbb{R}$ is nondegenerate if and only if the hessian $H_q : \mathbb{R}^m \times \mathbb{R}^m \to \mathbb{R}$ has nullity equal to zero.

Let q be a nondegenerate critical point for $f : M \to \mathbb{R}$. The **index of f at q** is the index of the hessian $H_q : \mathbb{R}^m \times \mathbb{R}^m \to \mathbb{R}$. This is well-defined, i.e., the index is independent of the choice of local coordinates. Moreover, the **Morse lemma** states that there is a coordinate chart $(\mathcal{U}, x_1, \ldots, x_m)$ centered at q such that

$$f|_{\mathcal{U}} = f(q) - (x_1)^2 - \ldots - (x_\lambda)^2 + (x_{\lambda+1})^2 + \ldots + (x_m)^2 \, ,$$

where λ is the index of f at q. In particular, nondegenerate critical points are necessarily isolated.

Let f be a Morse function on M. For $a \in \mathbb{R}$, let

$$M^a = f^{-1}(-\infty, a] = \{ p \in M \mid f(p) \le a \} \, .$$

---

[9] A point $q \in M$ is a **critical point** of f if $df_q = 0$. A critical point is **nondegenerate** if the **hessian matrix**

$$\left(\frac{\partial^2 f}{\partial x_i \partial x_j} \right)_q$$

is nonsingular, where the x_i's are local coordinates near q. (The condition that the hessian matrix is nonsingular is independent of the choice of coordinates.) The hessian matrix defines a symmetric bilinear function $H_q : \mathbb{R}^m \times \mathbb{R}^m \to \mathbb{R}$ given by inner product

$$(v, w) \longmapsto \left\langle v, \left(\frac{\partial^2 f}{\partial x_i \partial x_j} \right)_q w \right\rangle$$

and also called the **hessian of f at q** relative to the local coordinates x_i; the hessian is in fact the expression in coordinates of a natural bilinear form on the tangent space at q.

Theorem I.3.3. (Morse [40], Milnor [39])

(a) *Let $a < b$ and suppose that the set $f^{-1}[a, b]$, consisting of all $p \in M$ with $a \leq f(p) \leq b$, is compact, and contains no critical points of f. Then M^a is diffeomorphic to M^b. Furthermore, M^a is a deformation retract of M^b, so that the inclusion map $M^a \hookrightarrow M^b$ is a homotopy equivalence.*

(b) *Let q be a nondegenerate critical point with index λ and $f(q) = c$. Suppose that $f^{-1}[c - \varepsilon, c + \varepsilon]$ is compact, and contains no critical point of f other than q, for some $\varepsilon > 0$. Then, for all sufficiently small ε, the set $M^{c+\varepsilon}$ has the homotopy type of $M^{c-\varepsilon}$ with a λ-cell attached.*

(c) *If each set M^a is compact, then the manifold M has the homotopy type of a CW-complex with one cell of dimension λ for each critical point of index λ.*

A k-cell is simply a k-dimensional disk D^k, and it gets attached along its boundary S^{k-1}. Morse's original treatment did not include part (c) of Theorem I.3.3. Instead, his main results were phrased in terms of inequalities. Let $b_k(M) := \dim H_k(M)$ be the k-**th Betti number** of M. Let M be a compact manifold and f a Morse function on M. Let C_λ be the number of critical points of f with index λ.

Theorem I.3.4. (Morse inequalities [40])

(a)

$$b_\lambda(M) \leq C_\lambda ,$$

(b)

$$\sum_\lambda (-1)^\lambda b_\lambda(M) = \sum_\lambda (-1)^\lambda C_\lambda , \quad and$$

(c)

$$b_\lambda(M) - b_{\lambda-1}(M) + \ldots \pm b_0(M) \leq C_\lambda - C_{\lambda-1} + \ldots \pm C_0 .$$

A **perfect Morse function** is a Morse function for which the inequalities in the previous statement are equalities.

Corollary I.3.5. *If all critical points of a Morse function f have even index, then f is a perfect Morse function.*

I.3.3   Homology of Symplectic Toric Manifolds

Let $(M, \omega, \mathbb{T}^n, \mu)$ be a $2n$-dimensional symplectic toric manifold. Choose a suitably generic direction in $\mathbb{R}^n$ by picking a vector X whose components are independent over $\mathbb{Q}$. This condition ensures that:

- the one-dimensional subgroup, $T^X \subset \mathbb{T}^n$, generated by the vector X is dense in $\mathbb{T}^n$,

- X is nòt parallel to the facets of the moment polytope $\Delta := \mu(M)$, and

- the vertices of Δ have different projections along X.

Exercise 25

Check that the fixed points for the $\mathbb{T}^n$-action are exactly the fixed points of the action restricted to $\mathbb{T}^X$, that is, are the zeros of the vector field, $X^\#$ on M corresponding to the $\mathbb{T}^X$-action.

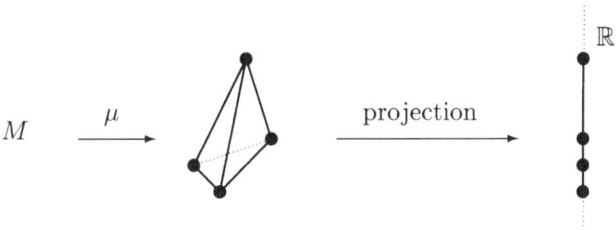

$$M \xrightarrow{\quad \mu \quad} \qquad \xrightarrow{\quad \text{projection} \quad} \qquad \mathbb{R}$$

Let $\mu^X := \langle \mu, X \rangle : M \to \mathbb{R}$ be the projection of μ along X. By definition of moment map, μ^X is a hamiltonian function for the vector field $X^\#$ generated by X. We conclude from the previous exercise that the critical points of μ^X are precisely the fixed points of the $\mathbb{T}^n$-action.

By Theorem I.3.2, if q is a fixed point for the $\mathbb{T}^n$-action, then there exists a chart $(\mathcal{U}, x_1, \ldots, x_n, y_1, \ldots, y_n)$ centered at q and weights $\lambda^{(1)}, \ldots, \lambda^{(n)} \in \mathbb{Z}^n$ such that

$$\mu^X\big|_{\mathcal{U}} = \langle \mu, X \rangle\big|_{\mathcal{U}} = \mu^X(q) - \frac{1}{2} \sum_{k=1}^{n} \langle \lambda^{(k)}, X \rangle (x_k^2 + y_k^2) \ .$$

Since the components of X are independent over $\mathbb{Q}$, all coefficients $\langle \lambda^{(k)}, X \rangle$ are nonzero, so q is a *nondegenerate* critical point of μ^X. Moreover, the index of q is twice the number of labels k such that $-\langle \lambda^{(k)}, X \rangle < 0$. But the $-\lambda^{(k)}$'s are precisely the edge vectors u_i which satisfy Delzant's conditions. Therefore, geometrically, the index of q can be read from the moment polytope Δ, by taking twice the number of edges whose inward-pointing edge vectors at $\mu(q)$ *point up relative to* X, that is, whose inner product with X is positive. In particular, μ^X is a perfect Morse function. By applying Corollary I.3.5 we conclude that:

Theorem I.3.6. *Let $X \in \mathbb{R}^n$ have components independent over $\mathbb{Q}$. The degree-2k homology group of the symplectic toric manifold $(M, \omega, \mathbb{T}, \mu)$ has dimension equal to the number of vertices of the moment polytope Δ where there are exactly k (primitive inward-pointing) edge vectors which point up relative to the projection along the X. All odd-degree homology groups of M are zero.*

Of course, by Poincaré duality (or by taking $-X$ instead of X), the words "point up" may be replaced by "point down".

Exercise 26
Let $(M, \omega, \mathbb{T}, \mu)$ be a symplectic toric manifold. What is the Euler characteristic of M?

I.3.4 Symplectic Blow-Up

Let L be the tautological line bundle over $\mathbb{P}^{n-1}$, that is,

$$L = \{([p], z) \mid p \in \mathbb{C}^n \setminus \{0\} , \ z = \lambda p \text{ for some } \lambda \in \mathbb{C}\}$$

with projection to $\mathbb{P}^{n-1}$ given by $([p], z) \mapsto [p]$. The fiber of L over the point $[p] \in \mathbb{P}^{n-1}$ is the complex line in $\mathbb{C}^n$ represented by that point.

Definition I.3.7. *The* **blow-up of $\mathbb{C}^n$ at the origin** *is the total space of the bundle L. The corresponding* **blow-down map** *is the map $\beta : L \to \mathbb{C}^n$ defined by $\beta([p], z) = z$.*

Notice that the total space of L may be decomposed as the disjoint union of two sets,

$$E := \{([p], 0) \mid p \in \mathbb{C}^n \setminus \{0\}\}$$

and

$$S := \{([p], z) \mid p \in \mathbb{C}^n \setminus \{0\} , \ z = \lambda p \text{ for some } \lambda \in \mathbb{C}^*\} .$$

The set E is called the **exceptional divisor**; it is diffeomorphic to $\mathbb{P}^{n-1}$ and gets mapped to the origin by β. On the other hand, the restriction of β to the complementary set S is a diffeomorphism onto $\mathbb{C}^n \setminus \{0\}$. Hence, we may regard L as being obtained from $\mathbb{C}^n$ by smoothly replacing the origin by a copy of $\mathbb{P}^{n-1}$.

There are actions of the unitary group $U(n)$ on all of these sets induced by the standard linear action on $\mathbb{C}^n$, and the map β is $U(n)$-equivariant.

Definition I.3.8. *A* **blow-up symplectic form** *on L is a $U(n)$-invariant symplectic form ω such that the difference $\omega - \beta^* \omega_0$ is compactly supported, where $\omega_0 = \frac{i}{2} \sum_{k=1}^{n} dz_k \wedge d\bar{z}_k$ is the standard symplectic form on $\mathbb{C}^n$.*

Two blow-up symplectic forms are called **equivalent** if one is the pullback of the other by a $U(n)$-equivariant diffeomorphism of L. Guillemin and Sternberg [26] have shown that two blow-up symplectic forms are equivalent if and only if they have equal restrictions to the exceptional divisor $E \subset L$.

Let Ω^ε ($\varepsilon > 0$) be the set of all blow-up symplectic forms on L whose restriction to the exceptional divisor $E \simeq \mathbb{P}^{n-1}$ is $\varepsilon \omega_{FS}$, where ω_{FS} is the Fubini-Study form on $\mathbb{P}^{n-1}$ described in Lecture 2. An ε-**blow-up** of $\mathbb{C}^n$ at the origin is a pair (L, ω) with $\omega \in \Omega^\varepsilon$.

Let (M, ω) be a $2n$-dimensional symplectic manifold. It is a consequence of the Darboux theorem that, for each point $q \in M$, there exists a chart $(\mathcal{U}, z_1, \ldots, z_n)$ centered at q and with image in $\mathbb{C}^n$ where

$$\omega|_{\mathcal{U}} = \frac{i}{2} \sum_{k=1}^{n} dz_k \wedge d\bar{z}_k \ .$$

It is shown in [26] that, for ε small enough, we can perform an ε-blow-up of M at q modeled on $\mathbb{C}^n$ at the origin, without changing the symplectic structure outside of a small neighborhood of q. The resulting manifold is then called an ε-**blow-up of** M **at** q.

Example. Let $\mathbb{P}(L \oplus \mathbb{C})$ be the $\mathbb{P}^1$-bundle over $\mathbb{P}^{n-1}$ obtained by projectivizing the direct sum of the tautological line bundle L with a trivial complex line bundle. Consider the map

$$\begin{array}{cccc} \beta : & \mathbb{P}(L \oplus \mathbb{C}) & \longrightarrow & \mathbb{P}^n \\ & ([p], [\lambda p : w]) & \longmapsto & [\lambda p : w] \ , \end{array}$$

where $[\lambda p : w]$ on the right represents a line in $\mathbb{C}^{n+1}$, forgetting that, for each $[p] \in \mathbb{P}^{n-1}$, that line sits in the 2-complex-dimensional subspace $L_{[p]} \oplus \mathbb{C} \subset \mathbb{C}^n \oplus \mathbb{C}$. Notice that β maps the *exceptional divisor*

$$E := \{([p], [0 : \ldots : 0 : 1]) \mid [p] \in \mathbb{P}^{n-1}\} \simeq \mathbb{P}^{n-1}$$

to the point $[0 : \ldots : 0 : 1] \in \mathbb{P}^n$, whereas β is a diffeomorphism on the complement

$$S := \{([p], [\lambda p : w]) \mid [p] \in \mathbb{P}^{n-1} \ , \ \lambda \in \mathbb{C}^* \ , \ w \in \mathbb{C}\} \simeq \mathbb{P}^n \setminus \{[0 : \ldots : 0 : 1]\} \ .$$

Therefore, we may regard $\mathbb{P}(L \oplus \mathbb{C})$ as being obtained from $\mathbb{P}^n$ by smoothly replacing the point $[0 : \ldots : 0 : 1]$ by a copy of $\mathbb{P}^{n-1}$. The space $\mathbb{P}(L \oplus \mathbb{C})$ is the blow-up of $\mathbb{P}^n$ at the point $[0 : \ldots : 0 : 1]$, and β is the corresponding blow-down map. The manifold $\mathbb{P}(L \oplus \mathbb{C})$ for $n = 2$ is known as the **first Hirzebruch surface**.

Exercise 27
Write a definition for *blow-up of a symplectic manifold along a complex submanifold* by considering the projectivization of the normal bundle to the submanifold.

Symplectic blow-up is due to Gromov according to the first printed exposition of this operation in [35].

I.3.5   Blow-Up of Toric Manifolds

Suppose that a compact Lie group acts on a symplectic manifold (M, ω) in a hamiltonian way, and that $q \in M$ is a fixed point for the G-action. Then, by Theorem I.3.1, there exists a Darboux chart $(\mathcal{U}, z_1, \ldots, z_n)$ centered at q which is G-equivariant with respect to a linear action of G on $\mathbb{C}^n$. Consider an ε-blow-up of M relative to this chart, for ε sufficiently small.

> **Exercise 28**
> Check that G acts on the blow-up in a hamiltonian way. Describe the moment map.

Let Δ be an n-dimensional Delzant polytope, and let $(M_\Delta, \omega_\Delta, \mathbb{T}^n, \mu_\Delta)$ be the associated symplectic toric manifold. The ε-blow-up of $(M_\Delta, \omega_\Delta)$ at a fixed point of the $\mathbb{T}^n$-action is a new symplectic toric manifold. What is the moment polytope Δ_ε corresponding to this new symplectic toric manifold?

Let q be a fixed point of the $\mathbb{T}^n$-action on $(M_\Delta, \omega_\Delta)$, and let $p = \mu_\Delta(q)$ be the corresponding vertex of Δ. (Cf. Exercise 16.) Let $u_1, \ldots, u_n$ be the primitive (inward-pointing) edge vectors at p, so that the rays $p + t u_i$, $t \geq 0$, form the edges of Δ at p.

Theorem I.3.9. *The ε-blow-up of $(M_\Delta, \omega_\Delta)$ at a fixed point q is the symplectic toric manifold associated to the polytope Δ_ε obtained from Δ by replacing the vertex p by the n vertices*

$$p + \varepsilon u_i , \quad i = 1, \ldots, n .$$

In other words, the moment polytope for the blow-up of $(M_\Delta, \omega_\Delta)$ at q is obtained from Δ by chopping off the corner corresponding to q, thus substituting the original set of vertices by the same set with the vertex corresponding to q replaced by exactly n new vertices:

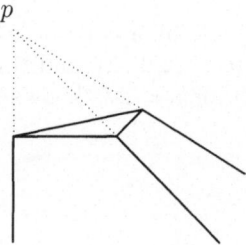

Proof. Exercise: Check that the new polytope is Delzant. We may view the ε-blow-up of $(M_\Delta, \omega_\Delta)$ as being obtained from M_Δ by smoothly replacing q by $(\mathbb{P}^{n-1}, \varepsilon \omega_{\mathrm{FS}})$. Compute the restriction of the moment map to this set. Recall Exercise 22.   □

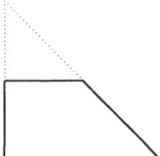

Example. The moment polytope for the standard $\mathbb{T}^2$-action on $(\mathbb{P}^2, \omega_{\text{FS}})$ is a right isosceles triangle Δ. If we blow-up $\mathbb{P}^2$ at $[0 : 0 : 1]$ we obtain a symplectic toric manifold associated to the trapezoid below.

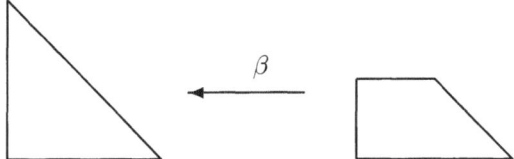

Exercise 29
Check that this manifold is the first Hirzebruch surface, defined in Section I.3.4.

Example. The following moment polytope corresponds to a toric manifold obtained by blowing-up $\mathbb{P}^2$ at the three fixed points:

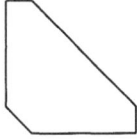

Exercise 30
The toric 4-manifold $\mathcal{H}_n$ corresponding to the polygon with vertices $(0,0)$, $(n+1,0)$, $(0,1)$ and $(1,1)$, for n a nonnegative integer, is called a **Hirzebruch surface**.

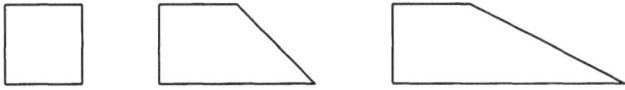

The manifold $\mathcal{H}_0$ is just a product of two spheres, whereas by the previous example $\mathcal{H}_1$ is a blow-up of $\mathbb{P}^2$ at a point.

   (a) Construct the manifold $\mathcal{H}_n$ by symplectic reduction of $\mathbb{C}^4$ with respect to an action of $(S^1)^2$.

   (b) Exhibit $\mathcal{H}_n$ as a $\mathbb{P}^1$-bundle over $\mathbb{P}^1$.

I.3.6 Symplectic Cutting

Let (M, ω) be a symplectic manifold where S^1 acts in a hamiltonian way, $\rho : S^1 \to$ $\mathrm{Diff}(M)$, with moment map $\mu : M \to \mathbb{R}$. Suppose that:

- M has a unique nondegenerate minimum at q where $\mu(q) = 0$, and

- for ε sufficiently small, S^1 acts freely on the level set $\mu^{-1}(\varepsilon)$.

Let $\mathbb{C}$ be equipped with the symplectic form $-idz \wedge d\bar{z}$. Then the action of S^1 on the product

$$\psi : S^1 \longrightarrow \mathrm{Diff}(M \times \mathbb{C}) , \qquad \psi_t(p, z) = (\rho_t(p), t \cdot z) ,$$

is hamiltonian with moment map

$$\phi : M \times \mathbb{C} \longrightarrow \mathbb{R} , \qquad \phi(p, z) = \mu(p) - |z|^2 .$$

Observe that S^1 acts freely on the ε-level of ϕ for ε small enough:

$$
\begin{aligned}
\phi^{-1}(\varepsilon) &= \{(p, z) \in M \times \mathbb{C} \mid \mu(p) - |z|^2 = \varepsilon\} \\
&= \{(p, 0) \in M \times \mathbb{C} \mid \mu(p) = \varepsilon\} \\
&\quad \cup \ \{(p, z) \in M \times \mathbb{C} \mid |z|^2 = \mu(p) - \varepsilon > 0\} .
\end{aligned}
$$

The reduced space is hence

$$\phi^{-1}(\varepsilon)/S^1 \simeq \mu^{-1}(\varepsilon)/S^1 \cup \{p \in M \mid \mu(p) > \varepsilon\} .$$

The open submanifold of M given by $\{p \in M \mid \mu(p) > \varepsilon\}$ embeds as an open dense submanifold into $\phi^{-1}(\varepsilon)/S^1$.

Exercise 31
Show that the reduced space $\phi^{-1}(\varepsilon)/S^1$ is the ε-blow-up of M at q.

This global description of blow-up for hamiltonian S^1-spaces is due to Lerman [31], as a particular instance of his *cutting* technique. **Symplectic cutting** is the application of symplectic reduction to the product of a hamiltonian S^1-space with the standard $\mathbb{C}$ as above, in a way that the reduced space for the original hamiltonian S^1-space embeds symplectically as a codimension 2 submanifold in a symplectic manifold.

As it is a local construction, the cutting operation may be more generally performed at a local minimum (or maximum) of the moment map μ.

There is a remaining S^1-action on the cut space $M^{\geq \varepsilon}_{\mathrm{cut}} := \phi^{-1}(\varepsilon)/S^1$ induced by

$$\tau : S^1 \longrightarrow \mathrm{Diff}(M \times \mathbb{C}) , \qquad \tau_t(p, z) = (\rho_t(p), z) .$$

In fact, τ is a hamiltonian S^1-action on $M \times \mathbb{C}$ which commutes with ψ, thus descends to an action $\tilde{\tau} : S^1 \to \mathrm{Diff}(M^{\geq \varepsilon}_{\mathrm{cut}})$.

Exercise 32
Show that $\tilde{\tau}$ is hamiltonian by describing a moment map.

Loosely speaking, the cutting technique provides a hamiltonian way to close the open manifold $\{p \in M \mid \mu(p) > \varepsilon\}$, by using the reduced space at level ε, $\mu^{-1}(\varepsilon)/S^1$. We may similarly close $\{p \in M \mid \mu(p) < \varepsilon\}$. The resulting hamiltonian S^1-spaces are called **cut spaces**, and denoted $M_{\text{cut}}^{\geq \varepsilon}$ and $M_{\text{cut}}^{\leq \varepsilon}$. Of course, if another group G acts on M in a hamiltonian way which commutes with the S^1-action, then the cut spaces are also hamiltonian G-spaces.

Chapter II

Algebraic Viewpoint

II.1 Toric Varieties

The goal of this lecture is to explain toric manifolds as a special class of projective varieties. The first five sections contain a crash course on notions and basic facts about algebraic varieties, mostly in order to fix notation. The combinatorial flavor of toric varieties is postponed until Lecture 5.

II.1.1 Affine Varieties

Let $\mathbb{C}[z_1, \ldots, z_n]$ be the algebra of polynomials in the n complex coordinate functions on $\mathbb{C}^n$. Throughout this lecture, we consider the **Zariski topology** on $\mathbb{C}^n$: a **(Zariski) closed set** in $\mathbb{C}^n$ is a set of common zeros of a finite number of polynomials from $\mathbb{C}[z_1, \ldots, z_n]$; naturally, the complement of a Zariski closed set is called a **(Zariski) open set**. The fact that any infinite intersection of closed sets is indeed a closed set follows from the stabilization property for sets given as zero sets of polynomials: any decreasing sequence of such sets $X_1 \supset X_2 \supset \ldots$ stabilizes, i.e., there exists an integer r such that $X_r = X_{r+1} = \ldots$. This is a restatement of Hilbert's basis theorem which says that any ideal in $\mathbb{C}[z_1, \ldots, z_n]$ is finitely generated; see, for instance, [50, §IV-1]. Notice that any nonempty open set is dense (i.e., its closure is the full space), hence the Zariski topology is not Hausdorff.

Definition II.1.1. *An* **affine variety** *is a nonempty closed set in a* $\mathbb{C}^n$.

The **Zariski topology on an affine variety** $X \subset \mathbb{C}^n$ is the topology on X which declares to be **(Zariski) open** (respectively, **closed**) every set which is the intersection of X with an open (resp., closed) subset of $\mathbb{C}^n$. An affine variety X is **irreducible** if it cannot be written as the union of two proper closed subsets. On an irreducible affine variety, any nonempty open subset is dense.

Example. The zero locus in $\mathbb{C}^2$ of the polynomial $z_1 z_2$ is not irreducible; its irreducible components are given by the lines $z_1 = 0$ and $z_2 = 0$.   ◇

> **Exercise 33**
>
> Let $X \subset \mathbb{C}^n$ be the affine variety defined as the zero locus of polynomials $p_1, \ldots, p_r \in \mathbb{C}[z_1, \ldots, z_n]$. Show that X is irreducible if and only if the ideal generated by $p_1, \ldots, p_r$ is *prime* (an ideal $I \subset \mathbb{C}[z_1, \ldots, z_n]$ is **prime** if whenever $u, v \in \mathbb{C}[z_1, \ldots, z_n]$ and $uv \in I$, then $u \in I$ or $v \in I$).

Let $X \subset \mathbb{C}^n$ be an affine variety presented as the zero locus of polynomials $p_1, \ldots, p_r \in \mathbb{C}[z_1, \ldots, z_n]$.

Definition II.1.2. *A* **regular function** *on X is a function $X \to \mathbb{C}$ which is the restriction to X of a polynomial function in $\mathbb{C}^n$. The ring of regular functions on X is denoted $\mathbb{C}[X]$.*

> **Exercise 34**
>
> Show that the ring $\mathbb{C}[X]$ is isomorphic to $\mathbb{C}[z_1, \ldots, z_n]/I$, where I is the ideal generated by $p_1, \ldots, p_r$.

Let $X \subset \mathbb{C}^n$ and $X' \subset \mathbb{C}^m$ be affine varieties.

Definition II.1.3. *A* **regular map** *from X to X' is a map $\varphi : X \to X'$ which is the restriction of a polynomial map $\mathbb{C}^n \to \mathbb{C}^m$.*

> **Exercise 35**
>
> Show that a map $\varphi : X \to X'$ is regular if and only if it pulls back regular functions on X' to regular functions on X.

Definition II.1.4. *An* **isomorphism** *from X to X' is a regular map $X \to X'$ which is invertible by a regular map. The symbol $\simeq$ indicates an isomorphism. The group of isomorphisms $X \to X$ is denoted by $\mathrm{Isom}(X)$. The affine varieties X and X' are* **isomorphic** *when there exists an isomorphism between them.*

> **Exercise 36**
>
> Show that X and X' are isomorphic if and only if the associated rings of regular functions, $\mathbb{C}[X]$ and $\mathbb{C}[X']$, are isomorphic.

Example. Consider the variety $X = \{z_i z_{n+i} = 1 , i = 1, \ldots, n\} \subset \mathbb{C}^{2n}$; for $n = 1$, this is the (complex) hyperbola with $\{z_1 = 0\}$ and $\{z_2 = 0\}$ as asymptotes. On X the functions $z_1, \ldots, z_n \in \mathbb{C}[X]$ are invertible by regular functions: $z_i^{-1} = z_{n+i}$.

Hence the ring of regular functions on X is the ring of Laurent polynomials in n variables:

$$\mathbb{C}[X] = \mathbb{C}[z_1, z_1^{-1}, \ldots, z_n, z_n^{-1}] \ .$$

The projection $\mathbb{C}^{2n} \to \mathbb{C}^n$ onto the first n components maps X isomorphically onto the n-**dimensional algebraic torus**

$$(\mathbb{C}^*)^n := (\mathbb{C} \setminus \{0\})^n = \mathbb{C}^n \setminus \text{ hyperplanes } z_i = 0 \ , i = 1, \ldots n \ .$$

Hence we have

$$
\begin{array}{ccc}
X & \overset{\simeq}{\longleftrightarrow} & (\mathbb{C}^*)^n \\
(z_1, \ldots, z_n, z_{n+1}, \ldots, z_{2n}) & \longmapsto & (z_1, \ldots, z_n) \\
(z_1, z_1^{-1}, \ldots, z_n, z_n^{-1}) & \longleftarrow & (z_1, \ldots, z_n) \ .
\end{array}
$$

This shows that the torus $(\mathbb{C}^*)^n$ is an affine variety. Note that the coordinates $z_1, \ldots, z_n$ are invertible by regular functions on $(\mathbb{C}^*)^n$, whereas they were not invertible by regular functions on $\mathbb{C}^n$. $\diamond$

II.1.2 Rational Maps on Affine Varieties

If $X \subset \mathbb{C}^n$ and $X' \subset \mathbb{C}^m$ are affine varieties, the symbol $X \dashrightarrow X'$ denotes a map defined on some open subset of X.

Example. If $p_1, p_2 \in \mathbb{C}[z_1, \ldots z_n]$, then the rational function defined on the set $\{p_2 \neq 0\}$ by $z \mapsto \frac{p_1(z)}{p_2(z)}$ is a map $\mathbb{C}^n \dashrightarrow \mathbb{C}$. $\diamond$

Let $X \subset \mathbb{C}^n$ be an irreducible affine variety.

Definition II.1.5. *A **rational function** on X is a map $X \dashrightarrow \mathbb{C}$ which is the restriction of a rational function on $\mathbb{C}^n$ whose denominator does not vanish identically on X. The ring of rational functions on X is denoted $\mathcal{O}_X$.*

Let $\mathbb{C}(X)$ denote the field of fractions of $\mathbb{C}[X]$, and let $\mathcal{I}_X$ be the ideal in $\mathcal{O}_X$ formed by the rational functions on $\mathbb{C}^n$ whose numerator vanishes identically on X.

Exercise 37
Show that the field $\mathbb{C}(X)$ is isomorphic to the quotient $\mathcal{O}_X/\mathcal{I}_X$.

An irreducible affine variety X is **normal** if its ring of regular functions $\mathbb{C}[X]$ is integrally closed in its field of fractions, that is, for any $f \in \mathbb{C}(X)$, if f satisfies an equation of the form

$$f^m + g_1 f^{m-1} + \ldots + g_m = 0$$

with coefficients $g_i \in \mathbb{C}[X]$, then $f \in \mathbb{C}[X]$.

Any smooth variety is normal and the set of singular points of a normal variety has codimension at least 2 [44].

Examples.

1. On the curve $X \subset \mathbb{C}^2$ defined by $y^2 = x^2 + x^3$, the rational function $t = \frac{y}{x} \in$ $\mathbb{C}(X)$ is integral over $\mathbb{C}[X]$ since $t^2 - 1 - x = 0$, but $t \notin \mathbb{C}[X]$, hence X is *not* normal.

2. The cone $X \subset \mathbb{C}^3$ given by $x^2 + y^2 = z^2$ is normal [44], though it has a singular point at the origin

Let $X \subset \mathbb{C}^n$ and $X' \subset \mathbb{C}^m$ be affine varieties.

Definition II.1.6. *A* **rational map** *from X to X' is a map $\varphi : X \dashrightarrow X'$ which is the restriction of a rational map $\mathbb{C}^n \dashrightarrow \mathbb{C}^m$ whose denominator does not vanish identically on X.*

> **Exercise 38**
> Show that a map $\varphi : X \dashrightarrow X'$ is rational if and only if it pulls back rational functions on X' to rational functions on X.

Definition II.1.7. *A* **birational equivalence** *from X to X' is a rational map $X \dashrightarrow X'$ which is invertible by a rational map. The affine varieties X and X' are* **birationally equivalent** *when there exists a birational equivalence between them.*

> **Exercise 39**
> Show that X and X' are birationally equivalent if and only if the associated fields of rational functions, $\mathbb{C}(X)$ and $\mathbb{C}(X')$, are isomorphic.

A **hypersurface in** $\mathbb{C}^n$ is the zero set of one polynomial in $\mathbb{C}[z_1, \ldots, z_n]$. Any affine variety is birationally equivalent to a hypersurface of some space $\mathbb{C}^m$; for a proof see, for example, [44].

II.1.3   Projective Varieties

We say that a polynomial $p \in \mathbb{C}[z_0, \ldots, z_n]$ vanishes at a point $[w_0 : \ldots : w_n] \in \mathbb{P}^n$ if $p(\lambda w_0, \ldots, \lambda w_n) = 0$ for all $\lambda \in \mathbb{C}^*$. Notice that this condition implies that each homogeneous component of p vanishes.

We consider the **Zariski topology** on $\mathbb{P}^n$: a **(Zariski) closed set** in $\mathbb{P}^n$ is a set of common zeros of a finite number of polynomials from $\mathbb{C}[z_0, \ldots, z_n]$; as usual, the complement of a Zariski closed set is called a **(Zariski) open set**. By considering homogeneous components, we may assume that each of those polynomials is homogeneous. A **hypersurface in** $\mathbb{P}^n$ is the zero set of one (reduced) homogeneous polynomial in $\mathbb{C}[z_0, \ldots, z_n] \setminus \{0\}$, the degree of which is called the **degree** of the hypersurface[1]. A hypersurface of degree $2, 3, 4, \ldots$ is traditionally called a **quadric** (except in $\mathbb{P}^2$ when it is called a **conic**), a **cubic**, a **quartic**, etc.[2]

Definition II.1.8. *A* **projective variety** *is a nonempty closed subset of some projective space* $\mathbb{P}^n$.

The **Zariski topology on a projective variety** $X \subset \mathbb{P}^n$ is the topology on X which declares to be **(Zariski) open** (respectively, **closed**) every set which is the intersection of X with an open (resp., closed) subset of $\mathbb{P}^n$. A projective variety X is **irreducible** if it cannot be written as the union of two proper closed subsets.

Examples.

1. The product of two projective spaces is a projective variety. This can be seen via the **Segre embedding**

$$S : \quad \mathbb{P}^n \times \mathbb{P}^m \quad \longrightarrow \quad \mathbb{P}^{nm+n+m}$$
$$([z], [w]) \quad \longmapsto \quad [z \otimes w] .$$

    The homogeneous coordinates of an image point $[z \otimes w]$ are

$$y_{ij} := z_i w_j , \quad i = 0, \ldots, n , \; j = 0, \ldots, m .$$

    The image of S is the set cut out by the system of equations

$$y_{ij} y_{k\ell} = y_{kj} y_{i\ell} , \qquad \begin{cases} i, k = 0, \ldots, n \\ j, \ell = 0, \ldots, m \end{cases}$$

    thus $S(\mathbb{P}^n \times \mathbb{P}^m) \simeq \mathbb{P}^n \times \mathbb{P}^m$ is a nonempty closed subset of $\mathbb{P}^{nm+n+m}$. In particular, $S(\mathbb{P}^1 \times \mathbb{P}^1)$ is the subset of points $[y_{00} : y_{01} : y_{10} : y_{11}] \in \mathbb{P}^3$ determined by the single quadratic equation

$$y_{00} y_{11} = y_{01} y_{10} ,$$

    hence $S(\mathbb{P}^1 \times \mathbb{P}^1)$ is a nondegenerate quadric in $\mathbb{P}^3$.

2. The set of lines in $\mathbb{P}^n$ through the point $[0 : \ldots : 0 : 1] \in \mathbb{P}^n$ (or, similarly, through any other point), is naturally identified with $\mathbb{P}^{n-1}$ via

$$\mathbb{P}^{n-1} \quad \longleftrightarrow \quad \{ \text{lines in } \mathbb{P}^n \text{ through } [0 : \ldots : 0 : 1] \}$$
$$[w] \quad \longleftrightarrow \quad L_{[w]} := \{ [\lambda w_0 : \ldots : \lambda w_{n-1} : \gamma] \mid (\lambda, \gamma) \in \mathbb{C}^2 \setminus \{(0, 0)\} \} .$$

---

[1] A polynomial is **reduced** if each of its irreducible factors has multiplicity 1. For some applications it is convenient to allow for nonreduced polynomials and then consider that some components of the hypersurface have multiplicities.

[2] A hypersurface of degree 1 is called a *line* when in $\mathbb{P}^2$, a *plane* when in $\mathbb{P}^3$, and a *hyperplane* in higher projective spaces.

The **blow-up of** $\mathbb{P}^n$ at $[0 : \ldots : 0 : 1]$, $B(\mathbb{P}^n, [0 : \ldots : 0 : 1])$, is the subset of $\mathbb{P}^n \times \mathbb{P}^{n-1}$ defined by the incidence relation

$$B(\mathbb{P}^n, [0 : \ldots : 0 : 1]) := \{([w], [z]) \in \mathbb{P}^{n-1} \times \mathbb{P}^n \mid [z] \in L_{[w]}\} .$$

This can be translated as the closed subset of $\mathbb{P}^n \times \mathbb{P}^{n-1}$ defined by the system of equations

$$z_i w_j = z_j w_i , \quad \forall\, 0 \leq i, j \leq n - 1 .$$

The Segre embedding exhibits $\mathbb{P}^n \times \mathbb{P}^{n-1}$ as a projective variety in $\mathbb{P}^{n^2+n-1}$, hence $B(\mathbb{P}^n, [0 : \ldots : 0 : 1])$ is a projective variety.

Since the point $[0 : \ldots : 0 : 1]$ belongs to any line through it, the set $B(\mathbb{P}^n, [0 : \ldots : 0 : 1])$ decomposes as the disjoint union of the so-called *exceptional divisor*,

$$E := \{([w], [0 : \ldots : 0 : 1]) \mid [w] \in \mathbb{P}^{n-1}\} \simeq \mathbb{P}^{n-1} ,$$

with

$$S := \{([w], [w : y]) \mid [w] \in \mathbb{P}^{n-1} , \ y \in \mathbb{C}\} \simeq \mathbb{P}^n \setminus \{[0 : \ldots : 0 : 1]\} .$$

$$\diamondsuit$$

Let $X \subset \mathbb{P}^n$ be a projective variety.

Definition II.1.9. *A regular function on X is a function $X \to \mathbb{C}$ which, locally near each point $x \in X$, may be written as a quotient of two homogeneous polynomials of the same degree such that the denominator does not vanish at x. The ring of regular functions on X is denoted $\mathbb{C}[X]$.*

Any regular function on $\mathbb{P}^n$ is constant. More generally, it can be shown that $\mathbb{C}[X] = \mathbb{C}$ whenever X is an irreducible projective variety [44]. Therefore, the ring $\mathbb{C}[X]$ will not give much information.

Let $X \subset \mathbb{P}^n$ and $X' \subset \mathbb{P}^m$ be projective varieties. Recall that complex projective space $\mathbb{P}^n$ comes equipped with $n + 1$ **affine neighborhoods** given by the standard charts ($k = 0. \ldots, n$):

$$\begin{aligned} V_k &:= \{[z_0 : \ldots : z_n] \mid z_k \neq 0\} \quad \simeq \quad \mathbb{C}^n \\ [z_0 : \ldots : z_n] &\mapsto \left(\tfrac{z_0}{z_k}, \ldots, \tfrac{z_{k-1}}{z_k}, \tfrac{z_{k+1}}{z_k}, \ldots, \tfrac{z_n}{z_k}\right) . \end{aligned}$$

Definition II.1.10. *A regular map from X to X' is a map $\varphi : X \to X'$ such that for each $x \in X$ there exists a neighborhood $\mathcal{U}$ of x and an affine neighborhood $\mathcal{V}$ of $\varphi(x)$ for which $\varphi(\mathcal{U}) \subset \mathcal{V}$ and $\varphi : \mathcal{U} \to \mathcal{V}$ is given by m regular functions.*

Exercise 40
Check that the regularity condition at a point $x \in X$ is independent of the choice of the affine neighborhood containing $f(x)$.

Example. A **Veronese embedding** of degree d is a map $V : \mathbb{P}^n \to \mathbb{P}^N$ where $N = \binom{n+d}{d} - 1$ and $[z_0 : \ldots : z_n]$ is mapped to the point with homogeneous coordinates given by the various monomials $z_0^{\lambda_0} \ldots z_n^{\lambda_n}$, the exponents $\lambda_0, \ldots, \lambda_n$ being nonnegative integers such that $\lambda_0 + \ldots + \lambda_n = d$.

Exercise 41
Check that the set of all homogeneous polynomials of degree d in $n+1$ variables $z_0, \ldots, z_n$ forms a vector space of dimension $\binom{n+d}{d}$.

In particular, if $n = 1$ and $d = 2$, the Veronese embedding is simply

$$V : \qquad \mathbb{P}^1 \quad \longrightarrow \quad \mathbb{P}^2$$
$$[z_0 : z_1] \quad \longmapsto \quad [z_0^2 : z_0 z_1 : z_1^2] \, .$$

Exercise 42
Check that V is a regular map.

The Veronese embedding of degree d allows to translate the study of some problems concerning hypersurfaces of degree d in $\mathbb{P}^n$ into the case of hyperplanes in $\mathbb{P}^N$. $\diamond$

Definition II.1.11. *An* **isomorphism** *from X to X' is a regular map $\varphi : X \to X'$ which is invertible by a regular map. The symbol $\simeq$ indicates an isomorphism. The group of isomorphisms $X \to X$ is denoted by* $\mathrm{Isom}(X)$. *The projective varieties X and X' are* **isomorphic** *when there exists an isomorphism between them.*

II.1.4 Rational Maps on Projective Varieties

If $X \subset \mathbb{P}^n$ and $X' \subset \mathbb{P}^m$ are projective varieties, the symbol $X \dashrightarrow X'$ still denotes a map defined on some open subset of X.

Let $X \subset \mathbb{P}^n$ be an irreducible projective variety.

Definition II.1.12. *A* **rational function** *on X is a function $\mathbb{P}^n \dashrightarrow \mathbb{C}$ whose restriction to each affine neighborhood is a rational function on $\mathbb{C}^n$ whose numerator and denominator have the same degree and whose denominator does not vanish identically on X. The ring of rational functions on X is denoted $\mathcal{O}_X$.*

Let $\mathbb{C}(X)$ denote the field of fractions of $\mathbb{C}[X]$, and let $\mathcal{I}_X$ be the ideal in $\mathcal{O}_X$ formed by the rational functions on $\mathbb{C}^n$ whose numerator vanishes identically on X.

Let $X \subset \mathbb{P}^n$ and $X' \subset \mathbb{P}^m$ be projective varieties.

Definition II.1.13. *A* **rational map from** X **to** $\mathbb{P}^m$ *is a map* $X \dashrightarrow \mathbb{P}^m$ *which is given in homogeneous coordinates for* $\mathbb{P}^m$ *by* $m+1$ *rational functions on* X*. A* **rational map from** X **to** X' *is the restriction to* X' *of a rational map* $\varphi : X \dashrightarrow \mathbb{P}^m$ *such that there is an open set* $\mathcal{U} \subset X$ *where* φ *is regular and* $\varphi(\mathcal{U}) \subset X'$.

Definition II.1.14. *A* **birational equivalence** *from* X *to* X' *is a rational map* $X \dashrightarrow X'$ *which is invertible by a rational map. The projective varieties* X *and* X' *are* **birationally equivalent** *when there exists a birational equivalence between them.*

II.1.5   Quasiprojective Varieties

The notions before make sense in the broader class of *quasiprojective varieties*, which encompasses both affine and projective varieties.

Definition II.1.15. *A* **quasiprojective variety** *is a nonempty open subset of a projective variety.*

Zariski topology, regular function, regular map, isomorphism, rational function, rational map and birational equivalence are defined for quasiprojective varieties analogously to projective varieties. For instance, a **(Zariski) closed set** of a quasiprojective variety is the intersection of the variety with a closed subset of projective space. From now on, the term *variety* (without specifying affine or projective) refers to a quasiprojective variety.

It is a fact that two irreducible varieties X and X' are birationally equivalent if and only if they contain isomorphic open subsets $\mathcal{U} \subset X$ and $\mathcal{U}' \subset X'$ [44].

Example. The tautological line bundle L defined in Section I.3.4 is a quasiprojective variety (yet not affine, nor projective). In fact, the inclusion of $\mathbb{C}^n$ in $\mathbb{P}^n$ as the open set of points $[z_0 : \ldots : z_n]$ with $z_n \neq 0$ induces an inclusion of the blow-up of $\mathbb{C}^n$ at the origin as an open subset of the projective variety $B(\mathbb{P}^n, [0 : \ldots : 0 : 1])$; cf. Section II.1.3.

The blow-down map $\beta : L \to \mathbb{C}^n$ is a birational equivalence in the category of quasiprojective varieties. $\Diamond$

Blowing-up at a point is a local operation which extends to any quasiprojective variety modeled on the blow-up of $\mathbb{C}^n$ at the origin or of $\mathbb{P}^n$ at $[0 : \ldots : 0 : 1]$.

Exercise 45
Check that $B(\mathbb{P}^n, [0 : \ldots : 0 : 1]) \simeq \mathbb{P}(L \oplus \mathbb{C})$, where $\mathbb{P}(L \oplus \mathbb{C})$ was discussed in Section I.3.4, and that the blow-down map $\beta : \mathbb{P}(L \oplus \mathbb{C}) \to \mathbb{P}^n$ is a birational equivalence in the category of projective varieties.

An irreducible projective variety X is **normal** if every point has a normal affine neighborhood.

A **normalization** of an irreducible variety X is an irreducible normal variety $\widetilde{X}$, together with a regular map $\nu : \widetilde{X} \to X$ which is a *finite* birational equivalence. A map $\varphi : X \to X'$ between varieties is **finite**, if any point $x \in X'$ has an affine neighborhood $\mathcal{V}$ such that the preimage $\mathcal{U} := \varphi^{-1}(\mathcal{V})$ is affine and the restriction $\varphi : \mathcal{U} \to \mathcal{V}$ is a finite map, that is every point has a finite number of preimages.

Any variety is a finite union of irreducible varieties [44]. If $X = \cup_i X_i$ presents X as a finite union of irreducible closed subsets and $X_i \not\subset X_j$ for all $i \neq j$, then the X_i's are called **irreducible components** of X.

One can define a normalization of an arbitrary variety X as a disjoint union of normalizations for each of the irreducible components of X.

II.1.6 Toric Varieties

The n-complex-dimensional **algebraic torus** $(\mathbb{C}^*)^n$ is a $2n$-dimensional Lie group under multiplication of complex numbers. The **weight lattice** of $(\mathbb{C}^*)^n$ is the lattice $\mathbb{Z}^n$. A **character** (or a *Laurent monomial*) of $(\mathbb{C}^*)^n$ is a group homomorphism $(\mathbb{C}^*)^n \to \mathbb{C}^*$.

There is a bijective correspondence between weights and characters of $(\mathbb{C}^*)^n$:

$$\lambda = (\lambda_1, \ldots, \lambda_n) \in \mathbb{Z}^n \longleftrightarrow \lambda : (\mathbb{C}^*)^n \to \mathbb{C}^*$$
$$w = (w_1, \ldots, w_n) \mapsto w^\lambda := w_1^{\lambda_1} \cdot \ldots \cdot w_n^{\lambda_n} .$$

An **action** of a torus $(\mathbb{C}^*)^n$ on a variety X is a group homomorphism $\psi : (\mathbb{C}^*)^n \to \mathrm{Isom}(X)$.

Definition II.1.16. *A* **toric variety** *is an irreducible variety[3] X equipped with an action of an algebraic torus having an open dense orbit.*

Definition II.1.17. *Two toric varieties are* **equivalent** *if there exists an equivariant isomorphism between them.*

Toric varieties are easy to construct, as the following examples illustrate.

---

[3]Toric varieties are usually required to be normal and normality is always the case for smooth varieties, which interest us mostly. However, our discussion does not require normality and is shortened without this assumption.

Examples.

1. Let $A = \{\lambda^{(1)}, \ldots, \lambda^{(k)}\}$ be a finite subset of $\mathbb{Z}^n$. Associated to A, there is an action of $(\mathbb{C}^*)^n$ on the projective space $\mathbb{P}^{k-1}$ defined by:

$$w \cdot [z_1 : \ldots : z_k] = [w^{\lambda^{(1)}} z_1 : \ldots : w^{\lambda^{(k)}} z_k], \qquad \text{for } w \in (\mathbb{C}^*)^n.$$

Let X_A be the closure of the $(\mathbb{C}^*)^n$-orbit through $[1 : \ldots : 1]$. Then X_A is a toric variety.

2. Let $A = \{\lambda^{(1)}, \ldots, \lambda^{(k)}\}$ be a finite subset of $\mathbb{Z}^n$. The action of $(\mathbb{C}^*)^n$ on the vector space $\mathbb{C}^k$ associated to A is defined by:

$$w \cdot (z_1, \ldots, z_k) = (w^{\lambda^{(1)}} z_1, \ldots, w^{\lambda^{(k)}} z_k), \qquad \text{for } w \in (\mathbb{C}^*)^n.$$

Let Y_A be the closure of the $(\mathbb{C}^*)^n$-orbit through $(1, \ldots, 1)$. Then Y_A is a toric variety.

3. Suppose that we have an action of a torus $(\mathbb{C}^*)^n$ on some variety V. Let $v \in V$. The closure of the $(\mathbb{C}^*)^n$-orbit through v is a toric variety.

$$\diamond$$

The important example X_A has the following particular instances.

Examples.

1. For a fixed positive integer d, let

$$\begin{aligned}
A_1 &= \{\lambda = (\lambda_0, \ldots, \lambda_n) \in \mathbb{Z}^{n+1} \mid \lambda_i \geq 0 \text{ all } i, \lambda_0 + \ldots + \lambda_n = d\} \\
&= \{\lambda^{(0)}, \ldots, \lambda^{(N)}\}
\end{aligned}$$

(which corresponds to the set of all Laurent monomials in n variables of degree d containing no negative powers). Then

$$X_{A_1} = \text{closure of } \{[w^{\lambda^{(0)}} : \ldots : w^{\lambda^{(N)}}] \mid w \in (\mathbb{C}^*)^{n+1}\} = V(\mathbb{P}^n) \simeq \mathbb{P}^n$$

where $V : \mathbb{P}^n \to \mathbb{P}^N$ is the Veronese embedding of degree d defined in Section II.1.3.

2. Similarly, for a fixed positive integer d, let

$$\begin{aligned}
A_2 &= \{\lambda = (\lambda_1, \ldots, \lambda_n) \in \mathbb{Z}^n \mid \lambda_i \geq 0 \text{ all } i, \lambda_1 + \ldots + \lambda_n \leq d\} \\
&\simeq \{\lambda = (\lambda_0, \ldots, \lambda_n) \in \mathbb{Z}^{n+1} \mid \lambda_i \geq 0 \text{ all } i, \lambda_0 + \ldots + \lambda_n = d\} \\
&= \{\lambda^{(0)}, \ldots, \lambda^{(N)}\}.
\end{aligned}$$

As before $X_{A_2} = V(\mathbb{P}^n) \simeq \mathbb{P}^n$. For $n = 2$ and $d = 3$, the set A_2 is:

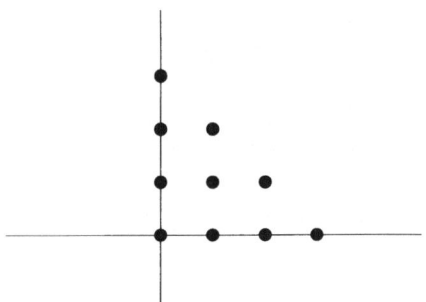

II.2   Classification

Recall that a toric variety is an irreducible quasiprojective variety equipped with an action of an algebraic torus having an open dense orbit. In this lecture we begin by reviewing the language of spectra used for classifying affine toric varieties. Arbitrary normal toric varieties are classified by combinatorial objects called fans.

II.2.1   Spectra

Let A be a finitely generated $\mathbb{C}$-algebra without zero divisors. An ideal I in A is **prime** if

$$u, v \in A \text{ and } uv \in I \implies u \in I \text{ or } v \in I.$$

An ideal I in A is **maximal** if $I \neq A$ and the only proper ideal in A containing I is I itself.

> **Exercise 46**
> Regard A simply as a commutative ring with unity. Show that the ideal I is prime if and only if the quotient ring A/I is an integral domain (i.e., A/I has no zero divisors), and that the ideal I is maximal if and only if A/I is a field.

> **Exercise 47**
> Check that every maximal ideal is prime. Give an example of a prime ideal which is not maximal.

> **Exercise 48**
> Let I be an ideal in A, and let $p : A \rightarrow A/I$ be the surjective ring homomorphism given by taking an element to its coset in the quotient ring A/I. Check that there exists a bijective correspondence between the ideals J of A which contain I, and the ideals $\overline{J}$ of A/I, given by $J = p^{-1}(\overline{J})$.

The **spectrum** of the algebra A is the set

$$\text{Spec } A := \{ \text{ prime ideals in } A \}$$

equipped with the **Zariski topology**, which declares to be **closed** a subset of Spec A consisting of all prime ideals containing some subset of A. The **maximal spectrum** of A is the set

$$\text{Spec}_{\mathrm{m}} A := \{ \text{ maximal ideals in } A \}$$

equipped with the **Zariski topology**, which declares to be **closed** a subset of $\text{Spec}_{\mathrm{m}} A$ consisting of all maximal ideals containing some subset of A.

Examples.

1. Let $A = \mathbb{C}[z_1, \ldots, z_n]$ and let $x = (x_1, \ldots, x_n) \in \mathbb{C}^n$. Associated to the point x there is a maximal ideal $I_x \subset A$ consisting of all polynomials which vanish at x, that is, I_x is the ideal generated by the monomials $z_1 - x_1, \ldots, z_n - x_n$:

$$I_x := \langle z_1 - x_1, \ldots, z_n - x_n \rangle .$$

By Hilbert's Nullstellensatz (see, for instance, [5] or [20]), any maximal ideal of $\mathbb{C}[z_1, \ldots, z_n]$ is of the form I_x for some $x \in \mathbb{C}^n$. Moreover, the correspondence

$$\mathbb{C}^n \simeq \mathrm{Spec}_m \underbrace{\mathbb{C}[z_1, \ldots, z_n]}_{\mathbb{C}[\mathbb{C}^n]}$$

is a homeomorphism for the Zariski topology, where we stress that the polynomial ring $\mathbb{C}[z_1, \ldots, z_n]$ is the ring of regular functions on $\mathbb{C}^n$.

> **Exercise 49**
> Show that there exists a bijective correspondence between $\mathrm{Spec}\ \mathbb{C}[z_1, \ldots, z_n]$ and the set of irreducible subvarieties in $\mathbb{C}^n$.
>
> **Hint:** Any prime ideal in $\mathbb{C}[z_1, \ldots, z_n]$ is finitely generated.

2. Let $A = \mathbb{C}[z_1, z_1^{-1}, \ldots, z_n, z_n^{-1}]$ and let $x = (x_1, \ldots, x_n) \in (\mathbb{C}^*)^n$. Associated to the point x there is the maximal ideal in A:

$$I_x := \langle z_1 - x_1, z_1^{-1} - x_1^{-1}, \ldots, z_n - x_n, z_n^{-1} - x_n^{-1} \rangle .$$

By observing that the ideal $\langle z_1 - x_1, z_1^{-1} - y_1, \ldots, \rangle$ is the full algebra A when $y_1 \neq x_1^{-1}$ since it contains $(z_1 - x_1)z_1^{-1} + x_1(z_1^{-1} - y_1) \in \mathbb{C}^*$, Hilbert's Nullstellensatz implies that any maximal ideal of $\mathbb{C}[z_1, z_1^{-1}, \ldots, z_n, z_n^{-1}]$ is of the form I_x. Moreover, the correspondence

$$(\mathbb{C}^*)^n \simeq \mathrm{Spec}_m \underbrace{\mathbb{C}[z_1, z_1^{-1}, \ldots, z_n, z_n^{-1}]}_{\mathbb{C}[(\mathbb{C}^*)^n]}$$

is a homeomorphism for the Zariski topology, where $\mathbb{C}[(\mathbb{C}^*)^n]$ is the ring of regular functions on $(\mathbb{C}^*)^n$; cf. Section II.1.1.

> **Exercise 50**
> Show that there exists a bijective correspondence between $\mathrm{Spec}\ \mathbb{C}[z_1, z_1^{-1}, \ldots, z_n, z_n^{-1}]$ and the set of irreducible subvarieties in $(\mathbb{C}^*)^n$.

Let X be an affine variety in $\mathbb{C}^n$ defined by polynomials $p_1, \ldots, p_r$ from $\mathbb{C}[z_1, \ldots, z_n]$. Let $I = \langle p_1, \ldots, p_r \rangle$ be the ideal generated by those polynomials.

Then

$$
\begin{aligned}
X &= \{\, x \in \mathbb{C}^n \mid p(x) = 0 \,,\ \forall p \in I \,\} \\
  &= \{\, x \in \mathbb{C}^n \mid I \subseteq I_x \,\} \\
  &\simeq \{\, \text{ideals } I_x \subset \mathbb{C}[z_1, \ldots, z_n] \mid I \subseteq I_x \,\} \\
  &= \{\, \text{maximal ideals } J \subset \mathbb{C}[z_1, \ldots, z_n] \mid I \subseteq J \,\} \\
  &= \{\, \text{maximal ideals in } \mathbb{C}[z_1, \ldots, z_n]/I \,\} \\
  &= \{\, \text{maximal ideals in } \mathbb{C}[X] \,\} \\
  &= \mathrm{Spec}_{\mathrm{m}}\, \mathbb{C}[X] \,.
\end{aligned}
$$

The correspondence $X \simeq \mathrm{Spec}_{\mathrm{m}}\, \mathbb{C}[X]$ is a homeomorphism for the Zariski topology.

More generally, if A is a finitely generated $\mathbb{C}$-algebra without zero divisors, the set $X_A := \mathrm{Spec}_{\mathrm{m}}\, A$ is called an **abstract affine variety**. While maximal ideals in A play the role of points in X_A, arbitrary prime ideals are thought of as irreducible subvarieties, by analogy with the ring of regular functions on an affine variety. The **dimension** of X_A is defined to be

$$
\dim X_A := \sup_{n \in \mathbb{Z}}\{\exists \text{ chain } I_0 \subset I_1 \subset \ldots \subset I_n \text{ of distinct prime ideals}\} \,.
$$

When $A = \mathbb{C}[X]$ is the ring of regular functions on an irreducible affine variety X, the dimension $\dim X_A$ coincides with the (complex) dimension $\dim X$ since both are equal to

$$
\sup_{n \in \mathbb{Z}}\{\exists \text{ chain } X_0 \subset X_1 \subset \ldots \subset X_n = X \text{ of distinct irreducible subvarieties}\} \,.
$$

II.2.2   Toric Varieties Associated to Semigroups

Let S be a commutative semigroup. Its **semigroup algebra** $\mathbb{C}[S]$ is the $\mathbb{C}$-algebra generated as a complex vector space by the symbols z^σ with $\sigma \in S$ and multiplication defined by the rule

$$
z^\sigma \cdot z^{\sigma'} = z^{\sigma + \sigma'} \,.
$$

In particular, generators σ_i for S as a semigroup yield generators z^{σ_i} for $\mathbb{C}[S]$ as a $\mathbb{C}$-algebra.

Examples.

1. If $S = (\mathbb{Z}_0^+)^n$, then $\mathbb{C}[S] = \mathbb{C}[z_1, \ldots, z_n]$ is the algebra of polynomials in n variables.

2. If $S = \mathbb{Z}^n$, then $\mathbb{C}[S] = \mathbb{C}[z_1, z_1^{-1}, \ldots, z_n, z_n^{-1}]$ is the algebra of Laurent polynomials in n variables.

Notice that in the previous two examples, the maximal spectrum of the semigroup algebras are toric varieties:

$$\mathrm{Spec}_{\mathrm{m}} \, \mathbb{C}[(\mathbb{Z}_0^+)^n] \simeq \mathbb{C}^n \quad \text{and} \quad \mathrm{Spec}_{\mathrm{m}} \, \mathbb{C}[\mathbb{Z}^n] \simeq (\mathbb{C}^*)^n \; .$$

In fact, we have the following general result:

Proposition II.2.1. *Let $S \subseteq \mathbb{Z}^n$ be a finitely generated semigroup. Then the maximal spectrum $\mathrm{Spec}_{\mathrm{m}} \, \mathbb{C}[S]$ is an affine toric variety.*

Proof. By shrinking the lattice $\mathbb{Z}^n$ if necessary, we may assume that S generates $\mathbb{Z}^n$ as an abelian group, in which case $\mathrm{Spec}_{\mathrm{m}} \, \mathbb{C}[S]$ has dimension n. The inclusion of semigroups $S \subseteq \mathbb{Z}^n$ and hence of semigroup algebras $\mathbb{C}[S] \subset \mathbb{C}[\mathbb{Z}^n]$ gives an embedding of the torus $(\mathbb{C}^*)^n$:

$$(\mathbb{C}^*)^n = \mathrm{Spec}_{\mathrm{m}} \, \mathbb{C}[\mathbb{Z}^n] \hookrightarrow \mathrm{Spec}_{\mathrm{m}} \, \mathbb{C}[S] \; .$$

Let $\mathcal{O}$ be the image of this embedding. The torus $(\mathbb{C}^*)^n$ acts on $\mathbb{C}[S]$ by

$$w \cdot z^\sigma := w^\sigma z^\sigma \; , \quad \text{for } w \in (\mathbb{C}^*)^n \text{ and } \sigma \in S \; .$$

This action induces an action of $(\mathbb{C}^*)^n$ on $\mathrm{Spec}_{\mathrm{m}} \, \mathbb{C}[S]$. By considering the dimension, the set $\mathcal{O} \simeq (\mathbb{C}^*)^n$ is an open orbit for this action. We conclude that its closure $\overline{\mathcal{O}}$ must be the full $\mathrm{Spec}_{\mathrm{m}} \, \mathbb{C}[S]$ and hence this is a toric variety. $\qquad \square$

Example. The complex curve in $\mathbb{C}^2$ with equation $y^k = x^{k+1}$ $(k = 1, 2, 3, \ldots)$ is an affine toric variety with $\mathbb{C}^*$-action given by

$$t \cdot (x, y) = (t^k x, t^{k+1} y) \; .$$

It may be obtained as $\mathrm{Spec}_{\mathrm{m}} \, \mathbb{C}[S]$ for the semigroup $S = \mathbb{Z}_0^+ \setminus \{1, 2, \ldots, k-1\}$ generated by $\{k, k+1, \ldots, 2k-1\}$. $\qquad \diamondsuit$

Exercise 51
Show that the variety in the previous example is not normal for $k > 1$, and that its normalization is the affine line X.

II.2.3 Classification of Affine Toric Varieties

Theorem II.2.2. (Classification of affine toric varieties) *Any affine toric variety is equivalent to one of the form $\mathrm{Spec}_{\mathrm{m}} \, \mathbb{C}[S]$ for some finitely generated semigroup $S \subset \mathbb{Z}^n$ $(n \geq 0)$.*

Proof. Let X be an affine toric variety for the torus $(\mathbb{C}^*)^m$. Let $\mathcal{O}$ be the open orbit for the $(\mathbb{C}^*)^m$-action on X. Then $\mathcal{O}$ may be identified with the quotient of $(\mathbb{C}^*)^m$ by the stabilizer of some point in $\mathcal{O}$. Since this quotient is itself a (possibly smaller dimensional) torus, we can regard $\mathcal{O}$ itself as a torus $(\mathbb{C}^*)^n$ acting on X. By irreducibility of X, we have an embedding $\mathbb{C}[X] \subset \mathbb{C}[\mathcal{O}] = \mathbb{C}[\mathbb{Z}^n]$. The subring $\mathbb{C}[X] \subset \mathbb{C}[\mathcal{O}]$ is $(\mathbb{C}^*)^n$-invariant with respect to the induced actions of $(\mathbb{C}^*)^n$ on $\mathbb{C}[X]$ and on $\mathbb{C}[(\mathbb{C}^*)^n]$. As a representation of an algebraic torus, the space $\mathbb{C}[X]$ decomposes into one-dimensional weight spaces. The weight spaces are generated by monomials as $\mathbb{C}$-algebras, hence the vector space $\mathbb{C}[X]$ itself is generated by monomials, i.e., it is a semigroup algebra. $\qquad\qquad\square$

Example. Recall the construction of the affine toric variety Y_A from a finite set $A = \{\lambda^{(1)}, \dots, \lambda^{(k)}\} \subset \mathbb{Z}^n$, described in Section II.1.6. By the previous theorem, we must have $Y_A \simeq \operatorname{Spec}_m \mathbb{C}[S]$ for some finitely generated semigroup $S \subset \mathbb{Z}^n$. It is not hard to see that S is the semigroup generated by A. In fact, the ring $\mathbb{C}[Y_A]$ of regular functions on Y_A is generated by the restrictions to Y_A of the coordinate functions on $\mathbb{C}^n$. Since Y_A is the closure of

$$\left\{ \left(z^{\lambda^{(1)}}, \dots, z^{\lambda^{(k)}} \right) \mid z \in (\mathbb{C}^*)^n \right\},$$

the ring $\mathbb{C}[Y_A]$ is generated by the monomials

$$z^{\lambda^{(1)}}, \dots, z^{\lambda^{(k)}},$$

i.e., is the semigroup algebra of the semigroup in $\mathbb{Z}^n$ generated by A. $\qquad\diamond$

Remark. The only *smooth* affine toric varieties are products of the form $(\mathbb{C}^*)^p \times \mathbb{C}^q$. This follows from the classification of affine toric varieties (Theorem II.2.2), the classification of normal toric varieties (Theorem II.2.10) and the study of conditions for smoothness (Exercise 53). See the remark at the end of Section II.3.4. $\qquad\diamond$

II.2.4 Fans

Definition II.2.3. *A* (**convex polyhedral**) **cone** *in $\mathbb{R}^n$ is a set of the form*

$$C = \{a_1 v_1 + \dots + a_r v_r \in \mathbb{R}^n \mid a_1, \dots, a_r \geq 0\}$$

for some finite set of vectors $v_1, \dots, v_r \in \mathbb{R}^n$, then called the **generators** *of the cone C. The cone C is* **rational** *if it admits a set of generators in $\mathbb{Z}^n$. The cone C is* **smooth** *if it admits a set of generators which is part of some $\mathbb{Z}$-basis of $\mathbb{Z}^n$.*

The **dimension** of a cone C is the dimension of the smallest $\mathbb{R}$-subspace containing C (which is the vector space $C + (-C)$).

Definition II.2.4. *The* **dual** *of a cone* $C \subset \mathbb{R}^n$ *is*

$$C^* := \{f \in (\mathbb{R}^n)^* \mid f(x) \geq 0 \ \forall x \in C\} \ .$$

Farkas' theorem states that the dual of a rational cone is a rational cone [21]. From the theory of convex sets [20], it follows that $(C^*)^* = C$.

A **supporting hyperplane** for a cone $C \subset \mathbb{R}^n$ is a hyperplane of the form

$$H_f := \{x \in \mathbb{R}^n \mid f(x) = 0\} \quad \text{for some } f \in C^* \setminus \{0\} \ .$$

A **face** of a cone $C \subset \mathbb{R}^n$ is either C itself (a nonproper face) or the intersection of C with any supporting hyperplane (proper faces). A face of a cone is itself a cone; indeed the face $C \cap H_f$ (with $f \in C^*$) is generated by those vectors v_i in a set of generators for C such that $f(v_i) = 0$.

> **Exercise 52**
>
> Show that a cone C has only finitely many faces and that any intersection of faces is also a face.
>
> **Hint:** The face $C \cap H_f$ is generated by those vectors v_i in a generating set for C such that $f(v_i) = 0$.

If 0 is a face of C, then C is called **strongly convex**; this is the case precisely when C contains no one-dimensional $\mathbb{R}$-subspaces, that is, when $C \cap (-C) = \{0\}$. If C is strongly convex, then its dual is n-dimensional (i.e., $C^* + (-C^*) = (\mathbb{R}^n)^*$), regardless of the dimension of C.

Let C be a rational cone in $\mathbb{R}^n$, and let C^* be its dual, also rational.

Lemma II.2.5. *The intersection* $C^* \cap (\mathbb{Z}^n)^*$ *is a finitely generated semigroup.*

Proof. Let $v_1, \ldots, v_r \in (\mathbb{Z}^n)^*$ be generators of C^* and let $K = \{\sum t_i v_i \mid 0 \leq t_i \leq 1\}$. The intersection $K \cap (\mathbb{Z}^n)^*$ is finite since K is compact. It suffices to show that $K \cap (\mathbb{Z}^n)^*$ generates S_C. For $v \in S_C$, write $v = \sum r_i v_i$ where $r_i \geq 0$, so $r_i = m_i + t_i$ with m_i a nonnegative integer and $0 \leq t_i \leq 1$. Then $v = \sum m_i v_i + \sum t_i v_i$ with each v_i and $\sum t_i v_i$ in $K \cap (\mathbb{Z}^n)^*$. $\square$

The finitely generated semigroup $C^* \cap (\mathbb{Z}^n)^*$ is denoted S_C.

Examples.

1. For the cone C given by the first octant in $\mathbb{Z}^n$, the semigroup S_C consists of elements in $(\mathbb{Z}^n)^*$ with all coordinates nonnegative, and is generated by $e_1^*, \ldots, e_n^*$, where $e_1 = (1, 0, \ldots, 0), \ldots, e_n = (0, \ldots, 0, 1) \in \mathbb{Z}^n$.

2. For the trivial strongly convex cone $C = \{0\}$ in $\mathbb{Z}^n$, the semigroup $S_C = (\mathbb{Z}^n)^*$ is generated by $e_1^*, -e_1^*, \ldots, e_n^*, -e_n^*$.

3. For the cone $C \subset \mathbb{Z}^2$ generated by e_2 and $e_1 - e_2$, the semigroup S_C is generated by e_1^* and $e_1^* + e_2^*$.

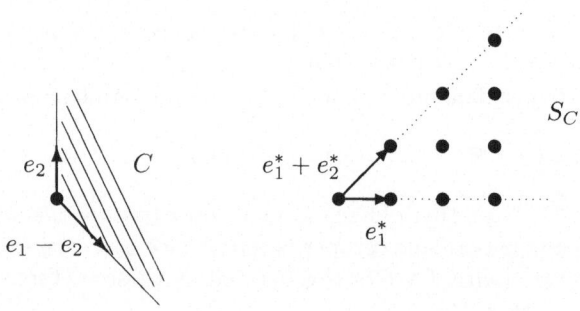

4. For the cone $C \subset \mathbb{Z}^2$ generated by e_2 and $2e_1 - e_2$, the semigroup S_C is generated by e_1^*, $e_1^* + e_2^*$ and $e_1^* + 2e_2^*$.

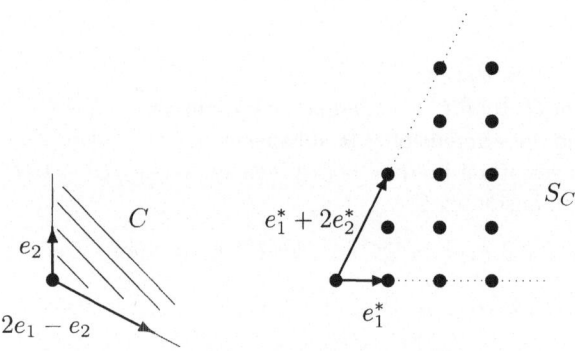

$\diamondsuit$

Corollary II.2.6. *For a rational cone $C \subset \mathbb{R}^n$, the affine variety $\mathrm{Spec}_{\mathrm{m}} \, \mathbb{C}[S_C]$ is a toric variety.*

Proof. This follows immediately from Proposition II.2.1 and Lemma II.2.5.   □

Examples.

1. For the cone C given by the first quadrant in $\mathbb{Z}^n$, the associated semigroup algebra is
$$\mathbb{C}[S_C] = \mathbb{C}[z_1, \ldots, z_n]$$
and the corresponding toric variety is
$$\mathrm{Spec}_{\mathrm{m}} \, \mathbb{C}[S_C] \simeq \mathbb{C}^n \, .$$

2. For the trivial cone $C = \{0\}$ in $\mathbb{Z}^n$, the associated semigroup algebra is

$$\mathbb{C}[S_C] = \mathbb{C}[z_1, z_1^{-1}, \ldots, z_n, z_n^{-1}]$$

and the corresponding toric variety is the torus

$$\mathrm{Spec}_{\mathrm{m}} \, \mathbb{C}[S_C] \simeq (\mathbb{C}^*)^n \ .$$

3. For the third cone in the previous list of examples, the associated semigroup algebra is
$$\mathbb{C}[S_C] = \mathbb{C}[z_1, z_1 z_2] \simeq \mathbb{C}[w_1, w_2]$$

and the corresponding toric variety is

$$\mathrm{Spec}_{\mathrm{m}} \, \mathbb{C}[S_C] \simeq \mathbb{C}^2 \ .$$

4. For the fourth cone in the previous list of examples, the associated semigroup algebra is

$$\mathbb{C}[S_C] = \mathbb{C}[z_1, z_1 z_2, z_1 z_2^2] \simeq \mathbb{C}[w_1, w_2, w_3]/\langle w_1 w_3 = w_2^2 \rangle$$

and the corresponding toric variety is the quadratic cone

$$\mathrm{Spec}_{\mathrm{m}} \, \mathbb{C}[S_C] \simeq \{(w_1, w_2, w_3) \in \mathbb{C}^3 \mid w_1 w_3 = w_2^2\}$$

which has an orbifold $\mathbb{Z}_2$ singularity at the origin. Note that, for the cone $C \subset \mathbb{Z}^2$ generated by e_2 and $e_1 - 2e_2$, the associated semigroup algebra is

$$\mathbb{C}[S_C] = \mathbb{C}[z_1, z_1^2 z_2] \simeq \mathbb{C}[w_1, w_2] \ ,$$

which corresponds to a smooth toric variety.

If we have an inclusion of rational cones $C \subset C'$, then $(C')^* \subset C^*$, hence $\mathbb{C}[S_{C'}]$ is a subalgebra of $\mathbb{C}[S_C]$. It follows that we get a map

$$\mathrm{Spec}_{\mathrm{m}} \, \mathbb{C}[S_C] \longrightarrow \mathrm{Spec}_{\mathrm{m}} \, \mathbb{C}[S_{C'}] \ ;$$

to see this, first notice that, since nontrivial $\mathbb{C}$-algebra homomorphisms $\mathbb{C}[S_C] \to \mathbb{C}$ are uniquely determined by their kernels which are exactly the maximal ideals of $\mathbb{C}[S_C]$, we obtain a bijection

$$\mathrm{Spec}_{\mathrm{m}} \, \mathbb{C}[S_C] \longleftrightarrow \mathrm{Hom}_{\mathbb{C}\text{-alg}}(\mathbb{C}[S_C], \mathbb{C}) \setminus \{0\} \ .$$

Thus, by restricting homomorphisms from $\mathbb{C}[S_C]$ to $\mathbb{C}[S_{C'}]$, we get the asserted map.

Lemma II.2.7. *Let C and C' be rational cones. If C is a face of C', then the induced map $\mathrm{Spec}_m\ \mathbb{C}[S_C] \to \mathrm{Spec}_m\ \mathbb{C}[S_{C'}]$ is an open injection for the Zariski topology.*

In other words, if C is a face of C', then $\mathrm{Spec}_m\ \mathbb{C}[S_C]$ is an open subset of $\mathrm{Spec}_m\ \mathbb{C}[S_{C'}]$.[4]

Proof. If C is a face of C', then (see, for instance, [21]) there is $v \in S_{C'}$ such that

$$S_C = S_{C'} + \mathbb{Z}_0^+(-v) \ ,$$

thus each element of $\mathbb{C}[S_C]$ may be written in the form $z^{\sigma - nv}$ for some $\sigma \in S_{C'}$ and $n \in \mathbb{Z}_0^+$. The map $\mathrm{Spec}_m\ \mathbb{C}[S_C] \to \mathrm{Spec}_m\ \mathbb{C}[S_{C'}]$ is injective, since if two $\mathbb{C}$-algebra homomorphisms $\mathbb{C}[S_C] \to \mathbb{C}$ coincide on $\mathbb{C}[S_{C'}]$, then they also coincide on elements z^{-nv}. The map $\mathrm{Spec}_m\ \mathbb{C}[S_C] \to \mathrm{Spec}_m\ \mathbb{C}[S_{C'}]$ misses exactly the maximal ideals in $\mathbb{C}[S_{C'}]$ containing the set $\{z^v\}$, since any $\mathbb{C}$-algebra homomorphism $\mathbb{C}[S_{C'}] \to \mathbb{C}$ which does not vanish on z^v extends to a nontrivial homomorphism $h : \mathbb{C}[S_C] \to \mathbb{C}$ where $h(z^{-v}) = h(z^v)^{-1}$, and if a $\mathbb{C}$-algebra homomorphism $\mathbb{C}[S_{C'}] \to \mathbb{C}$ vanishes on z^v, then any extension $h : \mathbb{C}[S_C] \to \mathbb{C}$ must vanish identically since $h(1) = h(z^v z^{-v}) = h(z^v)h(z^{-v}) = 0$. $\qquad\square$

Definition II.2.8. *A* **fan** *in $\mathbb{R}^n$ is a (nonempty) finite collection $\mathcal{F}$ of strongly convex rational cones such that*

- *every face of every cone $C \in \mathcal{F}$ belongs to $\mathcal{F}$,*

- *the intersection of any two cones from $\mathcal{F}$ is a face of both of them.*

The fan $\mathcal{F}$ is **smooth** *if all of its cones are smooth. The* **support** *of $\mathcal{F}$ is the union $|\mathcal{F}|$ of all cones from $\mathcal{F}$. The fan $\mathcal{F}$ is* **complete** *if $|\mathcal{F}|$ is the whole space.*

II.2.5   Toric Varieties Associated to Fans

Definition II.2.9. *The* **toric variety** *$X_{\mathcal{F}}$* **associated to a fan** *$\mathcal{F}$ in $\mathbb{R}^n$ is the result of gluing the affine toric varieties $X_C := \mathrm{Spec}_m\ \mathbb{C}[S_C]$ (for all $C \in \mathcal{F}$) by identifying X_C with the correponding Zariski open subset in $X_{C'}$ whenever C is a face of C'.*

Each affine chart X_C has a natural torus action defined as in the proof of Proposition II.2.1. Those actions are compatible under the identifications dictated by the face relations; hence there is a well-defined torus action on the variety $X_{\mathcal{F}}$. Moreover, $X_{\mathcal{F}}$ contains indeed an open dense orbit of $(\mathbb{C}^*)^n$ as the open set corresponding to the zero cone in $\mathcal{F}$: by strong convexity the zero cone is a face of every other cone, thus producing an open subset of each other affine piece, and the dual of the zero cone is the full set $(\mathbb{R}^n)^*$, so the corresponding algebra is $\mathbb{C}[z_1, z_1^{-1}, \ldots, z_n, z_n^{-1}]$ whose maximal spectrum is $(\mathbb{C}^*)^n$.

---

[4]This functorial property partly justifies working with cones and not their duals right away.

The variety $X_{\mathcal{F}}$ is normal since it is glued out of normal affine varieties and normality is a local property.

Exercise 53
Show that:

(a) The variety $X_{\mathcal{F}}$ is compact if and only if the fan $\mathcal{F}$ is complete.

(b) The variety $X_{\mathcal{F}}$ is smooth if and only if every cone from $\mathcal{F}$ is smooth.

As a consequence of the first part of the previous exercise, if $\mathcal{F}$ is a complete fan in $\mathbb{R}^n$, then $X_{\mathcal{F}}$ is a compactification of the torus $(\mathbb{C}^*)^n$.

Examples.

1. Consider the fan $\mathcal{F}$ consisting of the three cones

$$C = \{0\} , \quad C_0 = \mathbb{Z}_0^+(-e_1) \quad \text{and} \quad C_1 = \mathbb{Z}_0^+(e_1)$$

depicted below.

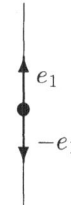

Each 1-dimensional cone represents the affine variety $\mathbb{C}$:

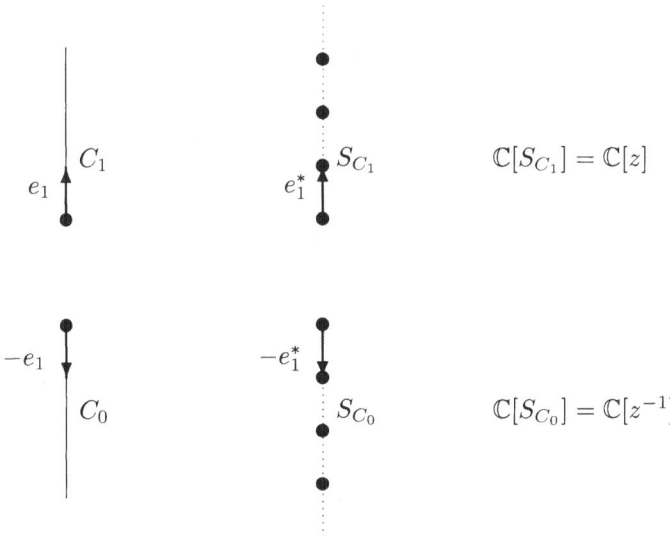

The gluing of these 1-dimensional charts is prescribed by the 0-dimensional cone C representing $\mathbb{C}^*$:

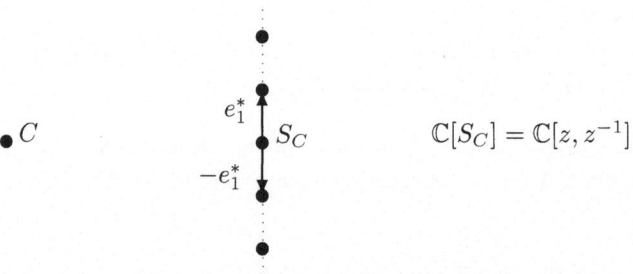

In X_{C_1}, the subset X_C corresponds to $\mathbb{C}^*_z := \{z \in \mathbb{C} \mid z \neq 0\}$, whereas in X_{C_0}, the subset X_C corresponds to $\mathbb{C}^*_{z^{-1}} := \{z^{-1} \in \mathbb{C} \mid z^{-1} \neq 0\}$. We can glue X_{C_1} to X_{C_0} along X_C by using the gluing map $z \mapsto z^{-1}$, thus producing $X_{\mathcal{F}} = \mathbb{P}^1$.

2. Consider the fan $\mathcal{F}$ consisting of seven cones (three 2-dimensional cones, three 1-dimensional cones and one 0-dimenisonal cone $C = \{0\}$), as sketched below. The shaded areas represent 2-dimensional cones.[5]

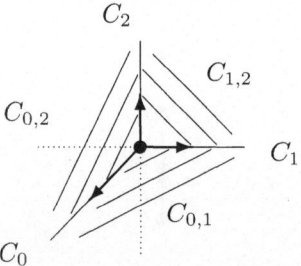

We will check that the toric variety $X_{\mathcal{F}}$ is $\mathbb{P}^2$. In fact, each 2-dimensional cone corresponds to an affine chart $\mathbb{C}^2$:

---

[5] If instead of the cones we considered their duals, the drawing would be messy with overlappings, hence the reason why we stick to this side of duality.

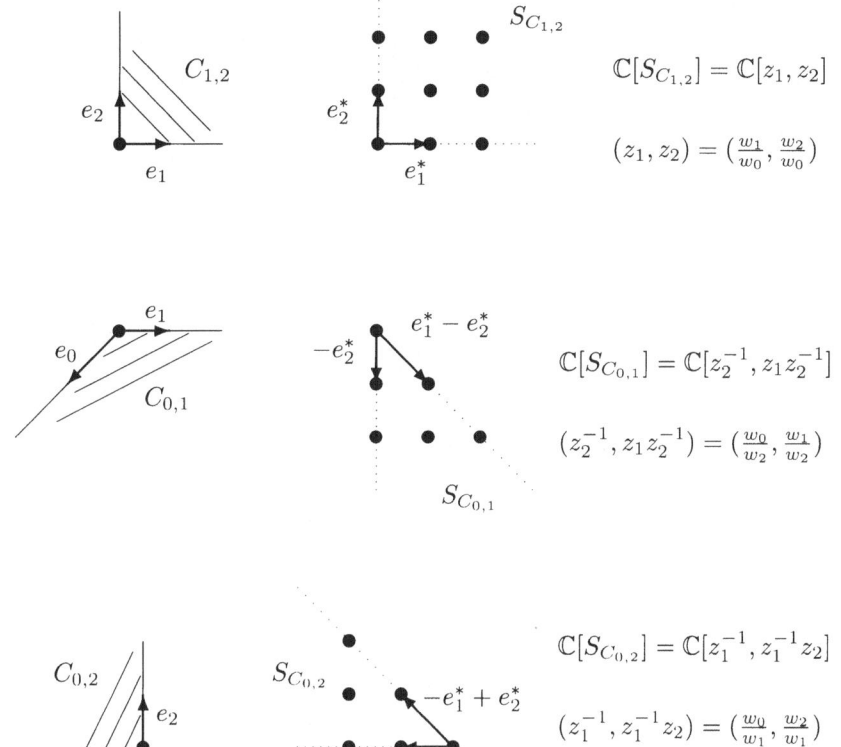

The expressions in terms of homogeneous coordinates $[w_0 : w_1 : w_2]$ on these three affine charts are written just to help keep track of the gluing maps below.[6]

The gluing of these affine charts along their intersections is prescribed by the 1-dimensional cones representing $\mathbb{C}^* \times \mathbb{C}$:

---

[6]The initial chosen identification

$$X_{C_{1,2}} \simeq \{[1 : z_1 : z_2] \mid (z_1, z_2) \in \mathbb{C}^2\}$$

determines the other two.

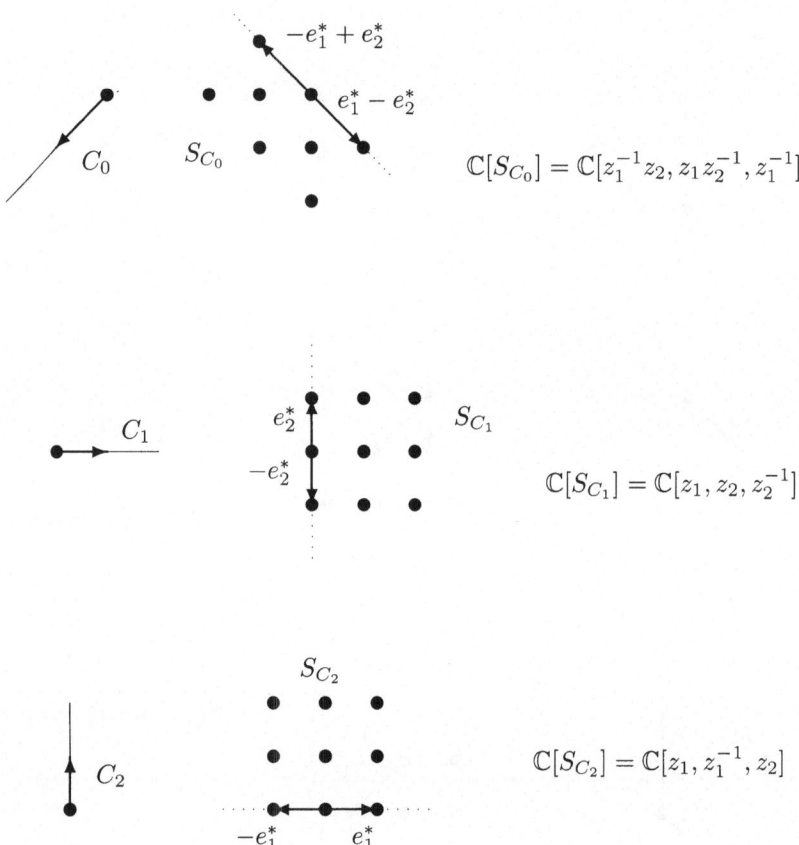

$$\mathbb{C}[S_{C_0}] = \mathbb{C}[z_1^{-1}z_2, z_1 z_2^{-1}, z_1^{-1}]$$

$$\mathbb{C}[S_{C_1}] = \mathbb{C}[z_1, z_2, z_2^{-1}]$$

$$\mathbb{C}[S_{C_2}] = \mathbb{C}[z_1, z_1^{-1}, z_2]$$

For instance, in $X_{C_{1,2}}$, the subset X_{C_1} is represented by $\mathbb{C}_{z_1} \times \mathbb{C}^*_{z_2}$ whereas in $X_{C_{0,1}}$, the subset X_{C_1} corresponds to $\mathbb{C}_{z_1 z_2^{-1}} \times \mathbb{C}^*_{z_2^{-1}}$. We can glue $X_{C_{1,2}}$ to $X_{C_{0,1}}$ along X_{C_1} by using the gluing map $(z_1, z_2) \mapsto (z_2^{-1}, z_1 z_2^{-1})$, to obtain $\mathbb{P}^2 \setminus \{[0:1:0]\}$.

3. Let $e_1 = (1, 0, \ldots, 0), \ldots, e_n = (0, \ldots, 0, 1)$ be the standard $\mathbb{Z}$-basis of $\mathbb{Z}^n$, and let $e_0 := -e_1 - \ldots - e_n$. Let $C_{i_1, \ldots, i_k}$ be the cone in $\mathbb{R}^n$ generated by the vectors $e_{i_1}, \ldots e_{i_k}$. Then the set

$$\mathcal{F} := \{C_{i_1, \ldots, i_k} \mid k \leq n, 0 \leq i_j \leq n\}\}$$

is a complete fan (where we include the trivial cone $C = \{0\}$.) In particular, for $n = 1, 2$ we get the fans encoding $\mathbb{P}^1$ and $\mathbb{P}^2$, respectively.

Exercise 54
Check that the toric variety associated to the fan $\mathcal{F}$ in the previous example
is projective space:

$$X_{\mathcal{F}} \simeq \mathbb{P}^n \ .$$

Exercise 55
Find the toric variety corresponding to the fan depicted below.

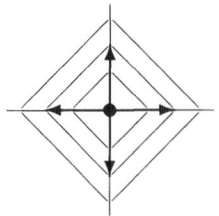

II.2.6 Classification of Normal Toric Varieties

Theorem II.2.10. (Classification of normal toric varieties) *Any normal toric
variety X is equivalent to a variety of the form $X_{\mathcal{F}}$ for some fan $\mathcal{F}$ in $\mathbb{R}^n$, where
n is the dimension of the torus acting on X. This fan is determined uniquely up
to a transformation from $\mathrm{GL}(n; \mathbb{Z})$.*

For a proof of Theorem II.2.10, see for instance [41]. Thanks to this theorem,
normal toric varieties are often defined in terms of fans.

Proposition II.2.11. *Let $\mathcal{F}$ be a fan in $\mathbb{R}^n$. Then the variety $X_{\mathcal{F}}$ has finitely many
orbits of the torus $(\mathbb{C}^*)^n$, and there is a natural bijection between the (nonempty)
cones in $\mathcal{F}$ and the $(\mathbb{C}^*)^n$-orbits in $X_{\mathcal{F}}$,*

$$\left\{ \begin{array}{c} cones \\ from \ \mathcal{F} \end{array} \right\} \longrightarrow \left\{ \begin{array}{c} (\mathbb{C}^*)^n\text{-}orbits \\ in \ X_{\mathcal{F}} \end{array} \right\}$$

$$C \longmapsto \mathcal{O}_C$$

*where the orbit $\mathcal{O}_C$ has dimension equal to the codimension of C. Moreover, for
cones $C, C' \in \mathcal{F}$ we have*

$$\mathcal{O}_{C'} \subset \overline{\mathcal{O}_C} \iff C \subset C' \ .$$

For a proof of Proposition II.2.11, see again [41]. In the next lecture we state
and prove the polytope analogue of this proposition.

II.3   Moment Polytopes

We have seen that normal toric varieties are classified by fans. Most interesting for our purposes are those fans dual to polytopes: a fan associated to a polytope defines an equivariantly projective toric variety. Moreover, a polytope encodes other geometric information such as a symplectic form and an equivariant complex line bundle. We will relate the dual languages of polytopes and fans, and review the link to the symplectic approach.

II.3.1   Equivariantly Projective Toric Varieties

Let X be a toric variety for a torus $(\mathbb{C}^*)^n$. We say that X is **equivariantly projective** if there exists a $(\mathbb{C}^*)^n$-equivariant embedding $X \hookrightarrow \mathbb{P}^k$ for some k and some action of $(\mathbb{C}^*)^n$ on $\mathbb{P}^k$.

Let $A = \{\lambda^{(1)}, \ldots, \lambda^{(k)}\}$ be a finite subset of $\mathbb{Z}^n$. The first example of a toric variety in Section II.1.6 is

$$X_A := \text{ closure of } \{[w^{\lambda^{(1)}} : \ldots : w^{\lambda^{(k)}}] \mid w \in (\mathbb{C}^*)^n\} \,,$$

that is, the closure of the $(\mathbb{C}^*)^n$-orbit through $[1 : \ldots : 1]$ for the action on $\mathbb{P}^{k-1}$ defined by the weights $\lambda^{(i)}$, $i = 1, \ldots, k$. The variety X_A is clearly *equivariantly* embedded in $\mathbb{P}^{k-1}$, i.e., the action of the torus on X_A extends to the whole ambient $\mathbb{P}^{k-1}$. The following theorem shows that, conversely, any equivariantly projective toric variety is equivalent to one of type X_A for some finite set $A \subset \mathbb{Z}^n$.

Theorem II.3.1. *Let X be a toric variety which is $(\mathbb{C}^*)^n$-equivariantly embedded in $\mathbb{P}^{\ell-1}$. Let Y be the minimal projective subspace in $\mathbb{P}^{\ell-1}$ containing X, and let $k-1$ be the (complex) dimension of Y. Then there exists a subset $A \subset \mathbb{Z}^n$ containing k elements and a $(\mathbb{C}^*)^n$-equivariant isomorphism $X_A \to X$ extending to an equivariant projective isomorphism $\mathbb{P}^{k-1} \to Y$.*

Proof. Any action of $(\mathbb{C}^*)^n$ on $\mathbb{P}^{\ell-1}$ by projective transformations can be lifted to a linear action on $\mathbb{C}^\ell$. Any linear action of $(\mathbb{C}^*)^n$ on $\mathbb{C}^\ell$ is diagonalizable, so in suitable coordinates it is given by a collection of weights $\lambda^{(i)} \in \mathbb{Z}^n$, $i = 1, \ldots, \ell$ such that $w \in (\mathbb{C}^*)^n$ acts as multiplication by

$$\begin{bmatrix} w^{\lambda^{(1)}} & & \\ & \ddots & \\ & & w^{\lambda^{(\ell)}} \end{bmatrix} .$$

Pick a point $z = [z_1 : \ldots : z_\ell] \in X$ lying on the open orbit of the torus. Let $A \subset \mathbb{Z}^n$ be the collection of those $\lambda^{(i)}$ for which $z_i \neq 0$. Then the minimal projective subspace of $\mathbb{P}^{\ell-1}$ containing X is

$$Y := \{[w_1 : \ldots : w_\ell] \in \mathbb{P}^{\ell-1} \mid z_i = 0 \Rightarrow w_i = 0\} \,.$$

The dimension of Y is $k - 1$ where $k = \#A$. We obtain a $(\mathbb{C}^*)^n$-equivariant isomorphism by collecting the nonzero coordinates:

$$
\begin{array}{ccc}
\mathbb{P}^{k-1} & \xrightarrow{\simeq} & Y \\
\cup & & \cup \\
X_A & \xrightarrow{\simeq} & X \, .
\end{array}
$$

$\square$

Example. Not all projective toric varieties are equivariantly projective. For instance, the nodal[7] cubic curve $X \subset \mathbb{P}^2$ given by the equation

$$
z_0 z_2^2 = z_1^3 - z_0 z_1^2
$$

is a (not normal) toric variety as its smooth part,

$$
X \setminus \{[1 : 0 : 0]\} \simeq \mathbb{C}^* \, ,
$$

is an open orbit for $\mathbb{C}^*$. For a reason why X does not admit an equivariant projective embedding see [22, p.169]. $\diamond$

II.3.2   Weight Polytopes

Consider a torus $(\mathbb{C}^*)^n$ acting linearly on a complex vector space V and consider the associated action on the projectivization $\mathbb{P}(V)$. Let v be a nonzero vector in V, and let $\overline{\mathcal{O}_v}$ be the closure in $\mathbb{P}(V)$ of the $(\mathbb{C}^*)^n$-orbit through $[v]$. By construction, $\overline{\mathcal{O}_v}$ is a toric variety equivariantly embedded in $\mathbb{P}(V)$. By Theorem II.3.1, the toric variety $\overline{\mathcal{O}_v}$ is equivalent to X_{A_v}, where A_v is the finite subset of $\mathbb{Z}^n$ prescribed as follows (adapting the proof of the previous theorem).

For a given a weight $\lambda \in \mathbb{Z}^n$ of $(\mathbb{C}^*)^n$, the λ-**weight space** of the $(\mathbb{C}^*)^n$-representation on V is the subspace

$$
V_\lambda = \{v \in V \mid w \cdot v = w^\lambda v \, , \ \forall w \in (\mathbb{C}^*)^n\} \, .
$$

The **weight space decomposition** of this representation is the isomorphism [11]

$$
V \simeq \bigoplus_{\lambda \in \mathbb{Z}^n} V_\lambda \, ;
$$

cf. Exercise 9. Given a vector $v \in V$, the **component of v of weight λ** is the component $v_\lambda \in V_\lambda$ of v in the weight space decomposition. Set

$$
A_v := \{\lambda \in \mathbb{Z}^n \mid v_\lambda \neq 0\} \, .
$$

---

[7]The adjective *nodal* refers to having no singularities other than ordinary double points.

Exercise 56

Show that the toric variety $\overline{\mathcal{O}}_v$ is equivalent to X_{A_v}.

Definition II.3.2. *The **weight polytope** P_v of the vector $v \in V \setminus \{0\}$ is the convex hull in $\mathbb{R}^n$ of the set A_v described in the previous paragraph.*

Example. Given a finite set $A \subset \mathbb{Z}^n$, its convex hull P in $\mathbb{R}^n$ is the weight polytope of the vector $v = (1, \ldots, 1)$ for which the closure of the orbit through $[v]$ is X_A. $\diamondsuit$

II.3.3   Orbit Decomposition

indexorbit decomposition

Let $A = \{\lambda^{(1)}, \ldots, \lambda^{(k)}\}$ be a finite subset of $\mathbb{Z}^n$. The toric variety associated to A is (cf. Section II.1.6)

$$X_A := \text{ closure of } \{[w^{\lambda^{(1)}} : \ldots : w^{\lambda^{(k)}}] \in \mathbb{P}^{k-1} \mid w \in (\mathbb{C}^*)^n\} \ .$$

Proposition II.3.3. *Let P be the convex hull in $\mathbb{R}^n$ of the set A above. Then there is a bijection between the (nonempty) faces of the polytope P and the $(\mathbb{C}^*)^n$-orbits in X_A given by*

$$\left\{ \begin{array}{c} \text{faces of} \\ \text{polytope } P \end{array} \right\} \longrightarrow \left\{ \begin{array}{c} (\mathbb{C}^*)^n\text{-orbits} \\ \text{in } X_A \end{array} \right\}$$

$$F \longmapsto X^0(F)$$

where

$$X^0(F) = \{[u] \in X_A \mid \forall \lambda : u_\lambda = 0 \Leftrightarrow \lambda \notin F\} \ .$$

The dimension of $X^0(F)$ is the dimension of F. Moreover, the closure $X(F)$ of $X^0(F)$ is equivalent to the toric variety $X_{A \cap F}$. If F and F' are two faces of P, then

$$X(F') \subset X(F) \iff F' \subset F \ .$$

Proof. For $u = (u_1, \ldots, u_k) \in \mathbb{C}^k$ such that $[u] := [u_1 : \ldots : u_k] \in X_A$, given $\lambda^{(i)} \in A$ we have that the $\lambda^{(i)}$-component of u is

$$u_{\lambda^{(i)}} = (0, \ldots, 0, u_i, 0, \ldots, 0) \ .$$

We first show that the correspondence is well-defined, i.e., that $X^0(F)$ is indeed an orbit. Of course, $X^0(F)$ is invariant as

$$[u] \in X^0(F) \Longrightarrow w \cdot [u] \in X^0(F) \ .$$

Besides, any two $[u], [u'] \in X^0(F)$ are related: to show that there exists indeed $w \in (\mathbb{C}^*)^n$ such that $u_\lambda = w^\lambda u_{\lambda'}$ for all $\lambda \in F$, notice that $u_\lambda = w_1^\lambda$ and $u_{\lambda'} = w_2^\lambda$ for some $w_1, w_2 \in (\mathbb{C}^*)^n$, hence just take $w = w_1 w_2^{-1}$.

To prove injectivity, we need to show that

$$F' \neq F \Longrightarrow X^0(F') \neq X^0(F) .$$

Without loss of generality, we assume that there exists $\lambda \in F'$ such that $\lambda \notin F$. Therefore, if $[u] \in X^0(F)$, then $u_\lambda = 0$, so that $[u] \notin X^0(F')$.

To prove surjectivity, we need to show that any $[u] \in X_A$ belongs to some $X^0(F)$. We introduce the notation

$$[v_F] := [v_1 : \ldots : v_k] \text{ such that } \begin{cases} v_i = 1 & \text{if } \lambda^{(i)} \in F \\ v_i = 0 & \text{if } \lambda^{(i)} \notin F ; \end{cases}$$

in particular we have that $[v_P] = [1 : \ldots : 1]$. By definition of X_A, any $[u] \in X_A$ is of the form

$$[u] = \lim_{z \to 0} f(z) \cdot [v_P]$$

for some analytic map of a punctured disk given by

$$f : \{z \in \mathbb{C}^* \mid |z| < \varepsilon\} \longrightarrow (\mathbb{C}^*)^n$$
$$f(z) = (c_1 z^{a_1} + \underbrace{\ldots}_{\text{h.d.}}, \ldots, c_n z^{a_n} + \underbrace{\ldots}_{\text{h.d.}}) ,$$

for some constants $c_i \in \mathbb{C}^*$, some exponents $a_i \in \mathbb{Z}$ and where the underbraced dots represent terms of higher degree. Given $a = (a_1, \ldots, a_n) \in \mathbb{R}^n$, let $f_a : \mathbb{R}^n \to \mathbb{R}$ be the linear function defined by inner product with a, $f_a(\lambda) := a \cdot \lambda$. Let

$$F_a := \{\lambda \in P \text{ where } f_a \text{ achieves its minimum } \} ;$$

the set F_a is a face of P, called the **supporting face of** f_a. For simplicity, suppose that $F_a \cap A = \{\lambda^{(1)}, \lambda^{(2)}\}$; the general case is just harder for notation. Let $c = (c_1, \ldots, c_n) \in (\mathbb{C}^*)^n$. Then

$$\begin{aligned} [u] &= \lim_{z \to 0} f(z) \cdot [v_P] \\[2mm] &= \lim_{z \to 0} [(f(z))^{\lambda^{(1)}} : \ldots : (f(z))^{\lambda^{(k)}}] \\[2mm] &= [c^{\lambda^{(1)}} + \underbrace{\ldots}_{\text{p.p.}} : c^{\lambda^{(2)}} + \underbrace{\ldots}_{\text{p.p.}} : \underbrace{\ldots}_{\text{p.p.}}] \\[2mm] &= \lim_{z \to 0} [c^{\lambda^{(1)}} : c^{\lambda^{(2)}} : 0 : \ldots : 0] \\[2mm] &= c \cdot [v_{F_a}] \in X^0(F_a) , \end{aligned}$$

where in the middle equality we have divided all homogeneous coordinates by $z^{a \cdot \lambda^{(1)}} = z^{a \cdot \lambda^{(2)}}$ and the underbraced dots represent terms with only positive powers of z.   □

II.3.4 Fans from Polytopes

Let $P \subset \mathbb{R}^n$ be a polytope, and let $f : \mathbb{R}^n \to \mathbb{R}$ be a linear function. We denote by $\mathrm{supp}_P f$ the supporting face of f in P, that is, the set of points in P where f achieves its minimum, as defined in the previous section.

Definition II.3.4. *Let F be a face of a polytope $P \subset \mathbb{R}^n$. The* **cone associated to F** *is the closure of the subset $C_{F,P} \subset (\mathbb{R}^n)^*$ consisting of all linear functions $f \in (\mathbb{R}^n)^*$ such that $\mathrm{supp}_P f = F$.*

Exercise 57
Show that $C_{F,P}$ is a convex cone, and that the collection of cones $C_{F,P}$ for all faces of P forms a complete fan.

Hint: Read the description in terms of the *dual polytope*, after the next definition, and translate P so that it contains the origin in its interior.

Definition II.3.5. *The* **fan of the polytope P** *is the collection $\mathcal{F}_P$ of the cones $C_{F,P}$ for all faces F of P.*

Suppose that the polytope P contains the origin in its interior. The fan of the polytope P coincides with the fan spanned by the faces of the *dual polytope*:

$$P^* := \{f \in (\mathbb{R}^n)^* \mid f(v) \geq -1 , \ \forall v \in P\} ,$$

that is, the collection of cones formed by the rays from the origin through the proper faces of P^*, plus the origin. For instance, the dual of a cube is an octahedron, so the fan of a cube has eight 3-dimensional triangular cones, together with all corresponding faces. Simple examples in $\mathbb{R}^2$ are:

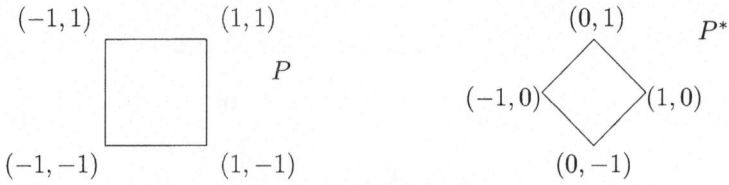

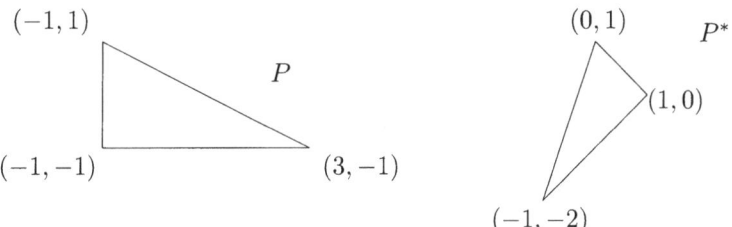

If P is rational, then (P^* is rational and) $\mathcal{F}_P$ is rational, and if P is smooth, then $\mathcal{F}_P$ is smooth.

Example. Let P be the polytope in the following picture.

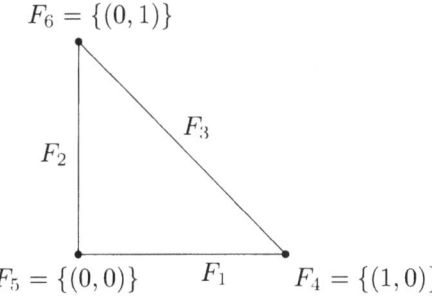

The fan of P is depicted below. The cone associated with the full polytope is the origin, whereas the cones associated with each of the facets F_1, F_2 and F_3, are half-lines, and the cones associated with each of the vertices F_4, F_5 and F_6 are two-dimensional (shaded regions).

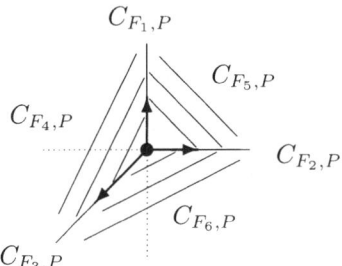

$\diamondsuit$

As in Section II.1.6, consider a torus $(\mathbb{C}^*)^n$ acting linearly on a vector space V and the associated action on the projectivization $\mathbb{P}(V)$. Let v be a nonzero vector in V, and let $\overline{\mathcal{O}_v}$ be the closure in $\mathbb{P}(V)$ of the $(\mathbb{C}^*)^n$-orbit through $[v]$. Then the toric variety $\overline{\mathcal{O}_v}$ is equivalent to X_{A_v}, where $A_v = \{\lambda \in \mathbb{Z}^n \mid v_\lambda \neq 0\}$.

Proposition II.3.6. *The fan of the toric variety $\overline{\mathcal{O}_v}$ equals the fan of the weight polytope P_v. In particular, $\dim \overline{\mathcal{O}_v} = \dim P_v$.*

For a proof of the previous proposition, see for instance [22, p.191] and [41].

Exercise 58
Check that, in a similar way, we can define a fan of a convex polyhedron, though in this case the fan may not be complete.

Remark. Let $\mathrm{Spec}_m \, \mathbb{C}[S]$ be the affine toric variety associated to the finitely generated semigroup $S \subseteq \mathbb{Z}^n$. We may assume that S generates $\mathbb{Z}^n$ as an abelian group. The variety $\mathrm{Spec}_m \, \mathbb{C}[S]$ is normal if and only if $S = P \cap \mathbb{Z}^n$ where P is the convex hull of S in $\mathbb{R}^n$. When $\mathrm{Spec}_m \, \mathbb{C}[S]$ is normal, its fan coincides with the fan of the convex polyhedron P. $\diamond$

II.3.5   Classes of Toric Varieties

Since all smooth varieties are normal, we restrict our attention to the universe of normal toric varieties, which are classified by fans. The *affine* ones correspond to fans consisting of the set of all faces of a single n-dimensional cone (see the remark at the end of the previous section). The *compact* ones correspond to complete fans. The *equivariantly projective* ones are necessarily of the form X_A for some set of the form $A = \mathbb{Z}^k \cap P$ where P is a polytope.[8] Since projective spaces are compact (or because fans of polytopes are complete), any equivariantly projective toric variety is compact.

Relation between these classes:

- Not all equivariantly projective normal toric varieties are smooth. To see a nonsmooth (i.e., singular) one just take the fan of a simple rational nonsmooth polytope. For instance, the triangle on the next page fails the smoothness condition at the top vertex.

---

[8]When $A \subset \mathbb{Z}^k$ is finite yet not of the form $\mathbb{Z}^k \cap P$ for some polytope P, the corresponding X_A is not normal, and its normalization is $X_{A'}$, where $A' = \mathbb{Z}^k \cap P'$ and P' is the convex hull of A.

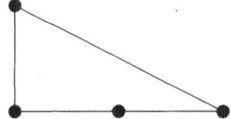

- Not all compact normal toric varieties are equivariantly projective, though in (complex) dimensions 1 and 2 this is always the case. Equivalently, not all complete fans come from polytopes in the sense of Definition II.3.5, though in dimensions 1 and 2 they do. There are plenty of complete fans in $\mathbb{R}^3$ which do not come from polytopes. For example, the collection of cones over the subdivision below of the boundary of the tetrahedron is not associated to any polytope [21, 41].

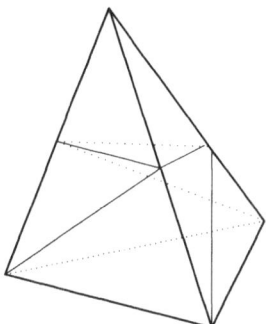

- Of course, not all normal toric varieties are compact – any affine toric variety is not compact; more generally, any fan which is not complete corresponds to a noncompact toric variety.

II.3.6 Symplectic vs. Algebraic

A **lattice polytope** in $\mathbb{R}^n$ is a polytope whose vertices belong to $\mathbb{Z}^n$.

Suppose that Δ is an n-dimensional polytope which is both Delzant and lattice. As a Delzant polytope, it is the moment polytope of a symplectic toric manifold $(M_\Delta, \omega_\Delta, \mathbb{T}^n, \mu_\Delta)$, by Delzant's construction.

On the other hand, consider the set of integral points in Δ:

$$A := \mathbb{Z}^n \cap \Delta = \{\lambda^{(1)}, \ldots, \lambda^{(k)}\} \,,$$

where $k := \#A$ is the number of such points. The convex hull of A is obviously Δ. Then the associated variety X_A is a toric variety for $(\mathbb{C}^*)^n$. The variety X_A is smooth and compact because the fan of the polytope Δ is smooth and complete, and X_A is connected because it is the closure of a $(\mathbb{C}^*)^n$-orbit. Moreover, by definition X_A is equivariantly embedded in $\mathbb{P}^{k-1}$,

$$i : X_A \hookrightarrow \mathbb{P}^{k-1} ,$$

and the restriction of the $(\mathbb{C}^*)^n$-action to its real subgroup

$$\mathbb{T}^n = \{(t_1, \ldots, t_n) \in (\mathbb{C}^*)^n \mid |t_i| = 1 \text{ for all } i \}$$

is effective, because the action of $(\mathbb{C}^*)^n$ was already effective.

Recall that projective spaces have canonical symplectic structures provided by the Fubini-Study forms. For later convenience, we equip $\mathbb{P}^{k-1}$ with the symplectic structure $-2\omega_{\mathrm{FS}}$. Since X_A is a *complex* submanifold of $\mathbb{P}^{k-1}$ and ω_{FS} is a Kähler form, we obtain that the restriction

$$\omega_A := i^*(-2\omega_{\mathrm{FS}})$$

is *nondegenerate*, hence a symplectic form on X_A. The structure ω_A is $\mathbb{T}^n$-invariant because ω_{FS} is $\mathbb{T}^n$-invariant.

We will check that the $\mathbb{T}^n$-action on X_A is in fact hamiltonian by exhibiting a moment map.

The action of $\mathbb{T}^n$ on $(\mathbb{C}^k, -2\omega_0)$ by

$$(t_1, \ldots, t_n) \cdot (z_1, \ldots, z_k) = (t^{\lambda^{(1)}} z_1, \ldots, t^{\lambda^{(k)}} z_k)$$

is hamiltonian with moment map

$$\widetilde{\mu} : \mathbb{C}^k \longrightarrow \mathbb{R}^n , \qquad \widetilde{\mu}(z_1, \ldots, z_k) = \sum_{j=1}^{k} \lambda^{(j)} |z_j|^2 .$$

The action of S^1 on $(\mathbb{C}^k, -2\omega_0)$ by diagonal multiplication

$$w \in S^1 \quad \longmapsto \quad \text{multiplication by} \quad \begin{bmatrix} w & & \\ & \ddots & \\ & & w \end{bmatrix}$$

is hamiltonian with moment map

$$\phi : \mathbb{C}^k \longrightarrow \mathbb{R} , \qquad \phi(z) = ||z||^2 - 1 .$$

The manifold $(\mathbb{P}^{k-1}, -2\omega_{\mathrm{FS}})$ is the symplectic reduction of $(\mathbb{C}^k, -2\omega_0)$ with respect to the S^1-action and the moment map ϕ. Since the S^1-action commutes with the action of $\mathbb{T}^n$ and preserves the moment map $\widetilde{\mu}$, we conclude that the $\mathbb{T}^n$-action and $\widetilde{\mu}$ descend to the quotient $(\mathbb{P}^{k-1}, -2\omega_{\mathrm{FS}})$. Therefore, the $\mathbb{T}^n$-action on $(\mathbb{P}^{k-1}, -2\omega_{\mathrm{FS}})$ by

$$(t_1, \ldots, t_n) \cdot [z_1 : \ldots : z_k] = [t^{\lambda^{(1)}} z_1 : \ldots : t^{\lambda^{(k)}} z_k]$$

is hamiltonian with moment map

$$\mu : \mathbb{P}^{k-1} \longrightarrow \mathbb{R}^n, \qquad \mu[z_1 : \ldots : z_k] = \frac{\sum\limits_{j=1}^k \lambda^{(j)} |z_j|^2}{\sum\limits_{j=1}^k |z_j|^2} .$$

The image of μ is the convex hull of A, i.e., is the polytope Δ.

As the symplectic submanifold (X_A, ω_A) is $\mathbb{T}^n$-invariant, the restriction of μ to X_A produces a moment map for the restricted action. We claim that the image of $\mu|_{X_A}$ is still Δ, so that the two constructions (Delzant's and the toric variety) yield equivalent symplectic toric manifolds.

Since $\mu(X_A)$ is a Delzant polytope, it suffices to show that each vertex of Δ is in $\mu(X_A)$. Let $\lambda^{(\ell)}$ be a vertex, and let $a = (a_1, \ldots, a_n) \in \mathbb{Z}^n$ be such that the restriction to Δ of the linear function $f_a : \mathbb{R}^n \to \mathbb{R}$, $f_a(\lambda) := a \cdot \lambda$, achieves its minimum in $\lambda^{(\ell)}$. Consider the map of a punctured disk to the torus

$$f : \{z \in \mathbb{C}^* \mid |z| < \varepsilon\} \longrightarrow (\mathbb{C}^*)^n$$
$$f(z) = (z^{a_1}, \ldots, z^{a_n}) .$$

By definition of X_A, we have that

$$\lim_{z \to 0} [(f(z))^{\lambda^{(1)}} : \ldots : (f(z))^{\lambda^{(k)}}] \in X_A .$$

Hence, by continuity of μ, we conclude that

$$\lambda^{(\ell)} = \lim_{z \to 0} \frac{\sum \lambda^{(j)} |f(z)^{\lambda^{(j)}}|^2}{\sum |f(z)^{\lambda^{(j)}}|^2} \in \mu(X_A) .$$

The coincidence of the two constructions allows to see that a symplectic toric manifold is Kähler, because it inherits a compatible invariant complex structure from its equivariant embeddings into projective spaces.

Remark. Not all toric varieties admit symplectic forms. A compact normal toric variety admits a symplectic form if and only if its fan comes from some polytope. Changing the cohomology class of the symplectic form corresponds to changing the lengths of the edges of the polytope. The size of the faces of a polytope cannot be recovered from the fan, where only the combinatorics of the faces is encoded. Hence, the fan does not give the cohomology class of the symplectic form. $\diamondsuit$

Exercise 59

Find the toric variety corresponding to the fan depicted below.

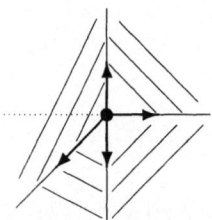

Bibliography

[1] Abraham, R., Marsden, J. E., *Foundations of Mechanics*, second edition, Addison-Wesley, Reading, 1978.

[2] Arnold, V., *Mathematical Methods of Classical Mechanics*, Graduate Texts in Mathematics **60**, Springer-Verlag, New York, 1978.

[3] Atiyah, M., Convexity and commuting Hamiltonians, *Bull. London Math. Soc.* **14** (1982), 1-15.

[4] Atiyah, M., Bott, R., The moment map and equivariant cohomology, *Topology* **23** (1984), 1-28.

[5] Atiyah, M., Macdonald, I., *Introduction to Commutative Algebra*, Addison-Wesley, Reading, 1969.

[6] Audin, M., *The Topology of Torus Actions on Symplectic Manifolds*, Progress in Mathematics **93**, Birkhäuser Verlag, Basel, 1991.

[7] Audin, M., *Spinning Tops. A Course on Integrable Systems*, Cambridge Studies in Advanced Mathematics **51**, Cambridge University Press, Cambridge, 1996.

[8] Batyrev, V., Mirror symmetry and toric geometry, Proceedings of the International Congress of Mathematicians, Vol. II (Berlin, 1998), *Doc. Math.* **1998**, Extra Vol. II, 239-248.

[9] Bott, R., Tu, L., *Differential Forms in Algebraic Topology*, Graduate Texts in Mathematics **82**, Springer-Verlag, New York-Berlin, 1982.

[10] Bredon, G., *Introduction to Compact Transformation Groups*, Pure and Applied Mathematics **46**, Academic Press, New York-London, 1972.

[11] Bröcker, T., tom Dieck, T., *Representations of Compact Lie Groups*, Graduate Texts in Mathematics **98**, Springer-Verlag, New York, 1985.

[12] Cannas da Silva, A., *Lectures on Symplectic Geometry*, Lecture Notes in Mathematics **1764**, Springer-Verlag, Berlin, 2001.

[13] Cox, D., Recent developments in toric geometry, *Algebraic Geometry – Santa Cruz 1995*, 389-436, Proc. Sympos. Pure Math. **62**, part 2, Amer. Math. Soc., Providence, 1997.

[14] Danilov, V., The geometry of toric varieties, *Uspekhi Mat. Nauk* **33** (1978), no. 2 (200), 85-134, 247, English translation: *Russian Math. Surveys* **33** (1978), no. 2, 97-154.

[15] Delzant, T., Hamiltoniens périodiques et images convexes de l'application moment, *Bull. Soc. Math. France* **116** (1988), 315-339.

[16] Demazure, M., Sous-groupes algébriques de rang maximum du groupe de Cremona, *Ann. Sci. École Norm. Sup.* (4) **3** (1970), 507-588.

[17] Duistermaat, J.J., On global action-angle coordinates, *Comm. Pure Appl. Math.* **33** (1980), 687-706.

[18] Duistermaat, J.J., Equivariant cohomology and stationary phase, *Symplectic Geometry and Quantization* (Sanda and Yokohama, 1993), edited by Maeda, Y., Omori, H. and Weinstein, A., 45-62, *Contemp. Math.* **179**, Amer. Math. Soc., Providence, 1994.

[19] Duistermaat, J.J., Heckman, G., On the variation in the cohomology of the symplectic form of the reduced phase space, *Invent. Math.* **69** (1982), 259-268; Addendum, *Invent. Math.* **72** (1983), 153-158.

[20] Ewald, G., *Combinatorial Convexity and Algebraic Geometry*, Graduate Texts in Mathematics **168**, Springer-Verlag, New York, 1996.

[21] Fulton, W., *Introduction to Toric Varieties*, Annals of Mathematics Studies **131**, Princeton University Press, Princeton, 1993.

[22] Gelfand, I., Kapranov, M., Zelevinsky, A., *Discriminants, Resultants, and Multidimensional Determinants*, Mathematics: Theory & Applications, Birkhäuser Boston, 1994.

[23] Griffiths, P., Harris, J., *Principles of Algebraic Geometry*, Chapter 0, reprint of the 1978 original, Wiley Classics Library, John Wiley & Sons, Inc., New York, 1994.

[24] Guillemin, V., *Moment Maps and Combinatorial Invariants of Hamiltonian T^n-spaces*, Progress in Mathematics **122**, Birkhäuser, Boston, 1994.

[25] Guillemin, V., Sternberg, S., Convexity properties of the moment mapping, *Invent. Math.* **67** (1982), 491-513.

[26] Guillemin, V., Sternberg, S., Birational equivalence in the symplectic category, *Invent. Math.* **97** (1989), 485-522.

[27] Guillemin, V., Sternberg, S., *Symplectic Techniques in Physics*, second edition, Cambridge University Press, Cambridge, 1990.

[28] Kempf, G., Knudsen, F., Mumford, D., Saint-Donat, B., *Toroidal Embeddings, I*, Lecture Notes in Mathematics **339**, Springer-Verlag, Berlin-New York, 1973.

[29] Kirwan, F., *Cohomology of Quotients in Symplectic and Algebraic Geometry*, Mathematical Notes **31**, Princeton University Press, Princeton, 1984.

[30] Kodaira, K., On the structure of compact complex analytic surfaces, I, *Amer. J. Math.* **86** (1964), 751-798.

[31] Lerman, E., Symplectic cuts, *Math. Res. Lett.* **2** (1995), 247-258.

[32] Lerman, E., Meinrenken, E., Tolman, S., Woodward, C., Nonabelian convexity by symplectic cuts, *Topology* **37** (1998), 245-259.

[33] Lerman, E., Tolman, S., Hamiltonian torus actions on symplectic orbifolds and toric varieties, *Trans. Amer. Math. Soc.* **349** (1997), 4201-4230.

[34] Marsden, J., Weinstein, A., Reduction of symplectic manifolds with symmetry, *Rep. Mathematical Phys.* **5** (1974), 121-130.

[35] McDuff, D., Examples of simply-connected symplectic non-Kählerian manifolds, *J. Differential Geom.* **20** (1984), 267-277.

[36] McDuff, D., Salamon, D., *Introduction to Symplectic Topology*, Oxford Mathematical Monographs, Oxford University Press, New York, 1995.

[37] Meyer, K., Symmetries and integrals in mechanics, *Dynamical Systems* (Proc. Sympos., Univ. Bahia, Salvador, 1971), 259-272. Academic Press, New York, 1973.

[38] Mikhalkin, G., Amoebas of algebraic varieties, survey for the Real Algebraic and Analytic Geometry Congress, Rennes, June 11-15, 2001, preprint available at http://xxx.lanl.gov/abs/math.AG/0108225.

[39] Milnor, J., *Morse Theory*, based on lecture notes by M. Spivak and R. Wells, Annals of Mathematics Studies **51**, Princeton University Press, Princeton, 1963.

[40] Morse, M., The foundations of a theory in the calculus of variations in the large, *Trans. Amer. Math. Soc.* **30** (1928), 213-274.

[41] Oda, T., *Convex Bodies and Algebraic Geometry – An Introduction to the Theory of Toric Varieties*, Ergebnisse der Mathematik und ihrer Grenzgebiete (3) **15**, Springer-Verlag, Berlin, 1988.

[42] Satake, I., On a generalization of the notion of manifold, *Proc. Nat. Acad. Sci. U.S.A.* **42** (1956), 359-363.

[43] Scott, P., The geometries of 3-manifolds, *Bull. London Math. Soc.* **15** (1983), 401-487.

[44] Shafarevich, I., *Basic Algebraic Geometry, 1, Varieties in Projective Space*, 2nd edition, translated from the 1988 russian edition and with notes by Miles Reid, Springer-Verlag, Berlin, 1994.

[45] Souriau, J.-M. , *Structure des Systèmes Dynamiques*, Maîtrises de Mathématiques, Dunod, Paris 1970.

[46] Spivak, M., *A Comprehensive Introduction to Differential Geometry*, Vol. I, second edition, Publish or Perish, Inc., Wilmington, 1979.

[47] Sturmfels, B., *Gröbner Bases and Convex Polytopes*, University Lecture Series **8**, Amer. Math. Soc., Providence, 1996.

[48] Weinstein, A., *Lectures on Symplectic Manifolds*, Regional Conference Series in Mathematics **29**, Amer. Math. Soc., Providence, 1977.

[49] Wells, R.O., *Differential Analysis on Complex Manifolds*, second edition, Graduate Texts in Mathematics **65**, Springer-Verlag, New York-Berlin, 1980.

[50] Zariski, O., Samuel, P., *Commutative Algebra*, Vol. 1, with the cooperation of I.S. Cohen, Graduate Texts in Mathematics **28**, Springer-Verlag, New York-Heidelberg-Berlin, 1975.

Index

Part C

Geodesic Flows and
Contact Toric Manifolds

Eugene Lerman

Foreword

The main theme of these notes is the topological study of contact toric manifolds, a relatively new class of manifolds that I find very interesting. A motivation for studying these manifolds comes from completely integrable systems $\{f_1, \ldots, f_n\}$ on punctured cotangent bundles where each function f_i is homogeneous of degree 1 (one can think of f_i's as symbols of first order pseudo-differential operators, but this is not essential). A punctured cotangent bundle is a symplectic cone whose base is naturally a contact manifold (this is explained in detail in Chapter I.2). This observation leads to studying completely integrable systems on contact manifolds, whatever those are.

The simplest (symplectic) completely integrable systems are the ones with global action-angle coordinates. The next simplest case is that of Hamiltonian torus actions. If the phase space is compact one ends up with (compact) symplectic toric manifolds. This is the theme of Ana Cannas's lectures delivered at the summer school. The corresponding case in the contact category is that of compact toric manifolds.

We will use the excuse of studying completely integrable geodesic flows with homogeneous integrals to introduce various ideas essential for the classification of contact and symplectic toric manifolds. More specifically we will discuss in these notes contact moment maps, slices for group actions, sheaves and Čech cohomology, orbifolds and Morse theory on orbifolds.

Chapter I

From toric integrable geodesic flows to contact toric manifolds

I.1   Introduction

We start with an innocuous sounding problem.

Problem   Consider the cotangent bundle of the n-torus minus the zero section $T^*\mathbb{T}^n \smallsetminus 0$. That is, consider the manifold

$$M = \{(q,p) \in \mathbb{R}^n/\mathbb{Z}^n \times \mathbb{R}^n \mid p \neq 0\}.$$

Suppose further that the torus $G = \mathbb{T}^n = \mathbb{R}^n/\mathbb{Z}^n$ acts on M effectively and preserves the standard symplectic form $\omega = \sum dp_i \wedge dq_i$ (i.e., the action is symplectic). Suppose further that the action of G commutes with dilations, i.e., the action ρ of $\mathbb{R}$ on M given by

$$\rho_\lambda(q,p) = (q, e^\lambda p).$$

Is the action of G necessarily free?

Remark I.1.   1. Recall that an action of the group G on a manifold M is **effective** if the only element of G that fixes all the points of M is the identity.

2. There is an "obvious" action of $\mathbb{T}^n$ on M which has the above properties and is free — it is the lift of left multiplication:

$$a \cdot (q,p) = (aq, p),$$

where $a \in \mathbb{T}^n$, $(q,p) \in M$. The issue is whether an **arbitrary** action of $\mathbb{T}^n$ which is effective, symplectic and commutes with dilations is necessarily free.

3. We will see later (Proposition I.13) that an action of a Lie group G on a punctured cotangent bundle which is symplectic and commutes with dilations is necessarily **Hamiltonian**. That is, there is a **moment map** $\Phi : M \to \mathfrak{g}^*$. Recall the definition of the moment map for a symplectic action of a Lie group G on a symplectic manifold (M, ω): for any vector X in the Lie algebra $\mathfrak{g}$ of G

$$d\langle\Phi, X\rangle = \omega(X_M, \cdot),$$

where $\langle\cdot,\cdot\rangle : \mathfrak{g}^* \times \mathfrak{g} \to \mathbb{R}$ is the canonical pairing and X_M is the vector field induced by the infinitesimal action of X: $X_M(x) = \frac{d}{dt}\big|_{t=0}(\exp tX) \cdot x$.

The problem is due to John Toth and Steve Zelditch [TZ]. The context is classical and quantum integrability of geodesic flows. Recall that for a manifold Q with a Riemannian metric g the corresponding **geodesic flow** is the flow on the cotangent bundle T^*Q of the Hamiltonian vector field X_h of the function $h = h_g \in C^\infty(T^*Q)$ which is the square root of the energy:

$$h_g(q, p) = (g_q^*(p, p))^{1/2}$$

for all $q \in Q$, $p \in T_q^*Q$. Here g^* denotes the inner product on the cotangent bundle $T^*Q \to Q$ dual to the inner product g; g^* is the so called dual metric. For an analyst the function h is the principal symbol of the square root of the Laplace operator $\sqrt{\Delta}$ defined by the metric g. A precise definition of $\sqrt{\Delta}$ will play no role in these notes. For a Riemannian geometer it's important that the integral curves of the vector field X_h project down to geodesics on the manifold Q. Toth and Zelditch were interested in the meaning of L^∞ boundedness of the L^2-normalized eigenfunctions of $\sqrt{\Delta}$. They observed that the question is easier if $\sqrt{\Delta}$ is quantum completely integrable. Again, it will not be important to us as to what that means precisely. What will matter is that quantum integrability of $\sqrt{\Delta}$ implies (classical) homogeneous complete integrability of the geodesic flow. Namely, it implies that there exist functions $f_1 = h, f_2, \ldots, f_n \in C^\infty(T^*Q \smallsetminus 0)$, $n = \dim Q$,[1] such that

1. the functions $f_1, \ldots, f_n$ are functionally independent on an open dense set $U \subset T^*Q \smallsetminus 0$, i.e., $df_1 \wedge \ldots \wedge df_n \neq 0$ on U;

2. the functions Poisson commute with each other: $\{f_i, f_j\} = 0$ for all $1 \leq i, j \leq n$;

3. the functions f_i are homogeneous of degree 1:

$$\rho_\lambda^* f_i = e^\lambda f_i$$

for all $\lambda \in \mathbb{R}$, where $\rho_\lambda : T^*Q \smallsetminus 0 \to T^*Q \smallsetminus 0$ again denotes the dilation $\rho_\lambda(q, p) = (q, e^\lambda p)$, $q \in Q$, $p \in T_q^*Q$.

---

[1] $T^*Q \smallsetminus 0$ denotes the punctured cotangent bundle of Q, that is, T^*Q with the zero section deleted.

Exercise I.2. Suppose f is a smooth function on the punctured cotangent bundle $T^*Q \smallsetminus 0$ of a manifold Q which is homogeneous of degree 1, i.e., $\rho_\lambda^* f = e^\lambda f$ for all dilations ρ_λ. Show that its Hamiltonian vector field X_f (relative to the standard symplectic structure on T^*Q) satisfies

$$d\rho_\lambda(X_f) = X_f \circ \rho_\lambda.$$

Conclude that the flow of X_h commutes with dilations.

Given a completely integrable system, we know that locally around any generic point there exist action-angle variables. Toth and Zelditch observed that things are considerably simpler if the action-angle variables are global. Then there exists an effective action of a torus $\mathbb{T}^n = \mathbb{R}^n/\mathbb{Z}^n$ on $T^*Q \smallsetminus 0$ preserving the function h and the symplectic form, and commuting with dilations ρ_λ. Toth and Zelditch proved (*op. cit.*):

Theorem I.3. *Suppose that the Lie group $G = \mathbb{T}^n$ acts effectively on $M = T^*\mathbb{T}^n \smallsetminus 0$, preserving the standard symplectic form and the function $h_g(q,p) = (g_q^*(p,p))^{1/2}$ for some metric g on $\mathbb{T}^n$. Suppose further the action commutes with dilations. If the action of G is free then the metric g is flat, that is,*

$$g = \sum g_{ij}\, dq_i \otimes dq_j$$

for some constants g_{ij} *(with $g_{ij} = g_{ji}$).*

The eigenfunctions of a flat metric Laplace operator on a torus are well understood.
    Let us now go back to the problem. The answer to the question is yes [LS]:

Theorem I.4. *Suppose the Lie group $G = \mathbb{T}^n = \mathbb{R}^n/\mathbb{Z}^n$ acts effectively on the punctured cotangent bundle $M = T^*\mathbb{T}^n \smallsetminus 0$ preserving the standard symplectic form and commuting with dilations $\rho_\lambda : M \to M$, $\rho_\lambda(q,p) = (q, e^\lambda p)$. Then the action of G is free.*

The main purpose of these notes is to explain why Theorem I.4 is true. I will now try to motivate the proof and to put it in a broader context. As was remarked previously (Remark I.1(3)), the action of the torus G on $M = T^*\mathbb{T}^n \smallsetminus 0$ is Hamiltonian. By the dimension count the manifold M together with its natural symplectic structure and the action of G is a symplectic toric manifold. Recall the definition.

Definition I.5. A **symplectic toric manifold** is a triple $(M, \omega, \Phi : M \to \mathfrak{g}^*)$ where M is a manifold, ω is a symplectic form on M, and Φ is a moment map for an effective Hamiltonian action of a torus G on (M, ω) satisfying $2 \dim G = \dim M$.

Compact symplectic toric manifolds are well understood thanks to a classification theorem of Delzant [D], which says that all such manifolds are classified by the images of the corresponding moment maps. Note that by the Atiyah-Guillemin-Sternberg convexity theorem [A, GS1], the images are convex rational polytopes.

Delzant proved that in the toric case the polytopes are simple[2] and additionally satisfy certain integrality conditions. Finally any simple polytope satisfying the integrality conditions occurs as an image of the moment map for a compact symplectic toric manifold.

Exercise I.6. Show that a Hamiltonian torus action on a *compact* symplectic manifold is never free. Hint: any smooth function on a compact manifold has a critical point (in fact it has at least two — a maximum and a minimum).

The manifold $M = T^*\mathbb{T}^n \smallsetminus 0$ we are interested in is not compact. Worse, we will see that the corresponding moment map $\Phi : M \to \mathfrak{g}^*$ is homogeneous (Proposition I.13):

$$\Phi(\rho_\lambda(m)) = e^\lambda \Phi(m)$$

for all $m \in M$, $\lambda \in \mathbb{R}$. This is a bit of bad news — Morse theory is an essential ingredient in the proof of the Atiyah-Guillemin-Sternberg convexity theorem and hence of Delzant's classification. Morse functions on a noncompact manifold which are not bounded either above or below are in practice impossible to work with.

On the other hand, because of homogeneity, the moment map descends to the quotient of M by the action of $\mathbb{R}$. The quotient is diffeomorphic to the co-sphere bundle $S^*\mathbb{T}^n := \{(q, p) \in T^*\mathbb{T}^n \mid g_q^*(p, p) = 1\}$ for some dual metric g^*. Since $S^*\mathbb{T}^n$ is odd-dimensional, it is not a symplectic manifold.[3]

It is also easy to see in an example that the map induced on $S^*\mathbb{T}^n$ by a homogeneous moment map on $T^*\mathbb{T}^n \smallsetminus 0$ behaves rather strangely if one is used to symplectic moment maps:

Example I.7. Consider the standard action of $G = \mathbb{T}^n$ on the cotangent bundle $T^*\mathbb{T}^n$, the lift of the left multiplication:

$$a \cdot (q, p) = (a \cdot q, p),$$

The corresponding moment map $\Phi : T^*G = G \times \mathfrak{g}^* \to \mathfrak{g}^*$ is given by $\Phi(q, p) = p$. Fix the standard metric on G and identify the dual of the Lie algebra $\mathfrak{g}^*$ with $\mathbb{R}^n$. Then the co-sphere bundle S^*G is $\mathbb{T}^n \times S^{n-1}$. The map $\Phi' = \Phi|_{S^*G} : S^*G \to \mathfrak{g}^*$ is also given by $\Phi'(q, p) = p$. Note that for any nonzero vector $X \in \mathfrak{g}$ the function $\langle \Phi', X \rangle$ has exactly two critical manifolds even though the action of G on S^*G is free!

Compare this with the symplectic situation where critical points of components of moment maps are points with nontrivial isotropy groups. What are we dealing with? We are dealing with moment maps for group actions on contact manifolds.

---

[2] A polytope in an n-dimensional real vector space is **simple** if there are exactly n edges meeting at each vertex. Equivalently, all the supporting hyperplanes are in general position. Thus a cube and a tetrahedron are simple and an octahedron is not.

[3] The manifold $S^*\mathbb{T}^n$ is contact. See next Chapter and the Appendix.

We finish the section by sketching a strategy for our proof of Theorem I.4. The effective action of the torus G on $M = T^*\mathbb{T}^n \smallsetminus 0$ descends to an effective action on the quotient $B = S^*\mathbb{T}^n$ of M by dilations. The action of G on B preserves a contact structure ξ (see Definition I.21 below) making (B, ξ) into a compact connected contact toric G-manifold (c.c.c.t.m., see Definition I.37). We then study all c.c.c.t.m.'s with a non-free torus actions and argue that none of them can have the homotopy type of $B = \mathbb{T}^n \times S^{n-1}$.

I.2 Symplectic cones and contact manifolds

In this section we define symplectic cones, contact forms and contact structures. Given a symplectic cone we show how to construct the corresponding contact manifold, and conversely, given a contact manifold we construct the corresponding symplectic cone. Thus symplectic manifolds and contact manifolds are "the same thing." Next we show that a symplectic action of a Lie group on a symplectic cone induces a contact action on the corresponding contact manifold. This will give us tools to set up a proof of Theorem I.4 as a study of contact toric manifolds. The material in this section is fairly well known. We now start by defining symplectic cones.

Definition I.8. A symplectic manifold (M, ω) is a **symplectic cone** if

- the manifold M is a principal $\mathbb{R}$ bundle over some manifold B, called the **base** of the cone, and

- the action of the real line $\mathbb{R}$ expands the symplectic form exponentially. That is, $\rho_\lambda^* \omega = e^\lambda \omega$, where ρ_λ denotes the diffeomorphism define by $\lambda \in \mathbb{R}$.

Definition I.9. Recall that a map $f : X \to Y$ between two topological spaces is **proper** if the preimage of a compact set under f is compact. An action of a Lie group G on a manifold M is **proper** if the map $G \times M \ni (g, m) \mapsto (g \cdot m, m) \in M \times M$ is proper.

By a theorem of Palais [P] the quotient M/G of a manifold M by a free proper action of a Lie group G is a manifold and the orbit map $M \to M/G$ makes M into a principal G bundle (see also Remark III.5 below). It follows that if a symplectic manifold (M, ω) has a complete vector field X with the following two properties:

1. the action of $\mathbb{R}$ induced by the flow of X is proper, and

2. the Lie derivative of the symplectic form ω with respect to the vector field X is again ω: $L_X \omega = \omega$,

then (M, ω) is a symplectic cone relative to the induced action of $\mathbb{R}$. This gives us an equivalent definition of a symplectic cone.

Definition I.10. A **symplectic cone** is a triple (M, ω, X) where M is a manifold, ω is a symplectic form on M, X is a vector field on M generating a proper action of the reals $\mathbb{R}$ such that $L_X \omega = \omega$.

Example I.11. Let (V, ω_V) be a symplectic vector space. The manifold $M = V \setminus \{0\}$ is a symplectic cone with the action of $\mathbb{R}$ given by $\rho_\lambda(v) = e^\lambda v$. Clearly $\rho_\lambda^* \omega_V = e^\lambda \omega_V$. The base is a sphere.

Example I.12. Let Q be a manifold. Denote the cotangent bundle of Q with the zero section deleted by $T^*Q \smallsetminus 0$. There is a natural free action of the reals $\mathbb{R}$ on the manifold $M := T^*Q \smallsetminus 0$ given by dilations $\rho_\lambda(q, p) = (q, e^\lambda p)$. It expands the standard symplectic form on the cotangent bundle exponentially. Thus $T^*Q \smallsetminus 0$ is naturally a symplectic cone. The base is the co-sphere bundle S^*Q.

Proposition I.13. *Suppose (M, ω, X) is a symplectic cone and suppose a Lie group G acts on M preserving the symplectic form ω and the expanding vector field X. Then the action of G on the symplectic manifold (M, ω) is Hamiltonian. Moreover we may choose the moment map $\Phi : M \to \mathfrak{g}^*$ to be homogeneous of degree 1, i.e.,*

$$\Phi(\rho_\lambda(m)) = e^\lambda \Phi(m)$$

for all $\lambda \in \mathbb{R}$ and $m \in M$. Here ρ_λ denotes the action of $\mathbb{R}$ generated by X, that is, the time λ flow of X.

Proof. Note first that since $L_X \omega = \omega$ and since $d\omega = 0$, it follows from Cartan's formula $(L_X \omega = \iota(X)d\omega + d\iota(X)\omega)$ that $d(\iota(X)\omega) = \omega$. Since the action of G preserves X and ω, it preserves the contraction $\iota(X)\omega$. Therefore for any vector A in the Lie algebra $\mathfrak{g}$ of G we have $L_{A_M}(\iota(X)\omega) = 0$, where A_M as before denotes the vector field on M induced by A. Therefore

$$0 = d\iota(A_M)\iota(X)\omega + \iota(A_M)d(\iota(X)\omega) = d(\omega(X, A_M)) + \iota(A_M)\omega,$$

and consequently
$$\iota(A_M)\omega = d(\omega(A_M, X)).$$

We conclude that the map $\Phi : M \to \mathfrak{g}^*$ defined by

$$\langle \Phi(m), A \rangle = \omega_m(X(m), A_M(m))$$

is a moment map for the action of G on (M, ω). $\qquad\square$

Exercise I.14. Suppose a Lie group G acts on a manifold M preserving a 1-form β. Define the β-**moment map** $\Psi_\beta : M \to \mathfrak{g}^*$ by

$$\langle \Psi_\beta(m), A \rangle = \beta_m(A_M(m))$$

for all $A \in \mathfrak{g}$ and all $m \in M$. Here as usual A_M denotes the vector field induced by A on M.

Show that Ψ_β is G-equivariant, that is, show that for any $a \in G$ and any $m \in M$

$$\Psi_\beta(a \cdot m) = Ad^\dagger(a)\Psi_\beta(m),$$

where $Ad^\dagger : G \to \mathrm{GL}(\mathfrak{g}^*)$ denotes the coadjoint representation. Conclude that the map Φ defined in Proposition I.13 is equivariant.

Definition I.15. A 1-form α on a manifold B is a **contact form** if the following two conditions hold:

1. $\alpha_b \neq 0$ for all points $b \in B$. Hence $\xi := \ker \alpha = \{(b, v) \in TB \mid \alpha_b(v) = 0\}$ is a vector subbundle of the tangent bundle TB.

2. $d\alpha|_\xi$ is a symplectic structure on the vector bundle $\xi \to B$ (i.e. $d\alpha_b|_{\xi_b}$ is nondegenerate).

Remark I.16.   1. If $\xi \to B$ is a symplectic vector bundle, then the dimension of its fibers is necessarily even. Hence if a manifold B has a contact form, then B is odd-dimensional.

2. A 1-form α on $2n + 1$ dimensional manifold B is contact if and only if the form $\alpha \wedge (d\alpha)^n$ is never zero, i.e., it is a volume form. [Prove this].

Example I.17. The 1-form $\alpha = dz + x\,dy$ on $\mathbb{R}^3$ is a contact form: $\alpha \wedge d\alpha = dz \wedge dx \wedge dy$.

Example I.18. Let $B = \mathbb{R} \times \mathbb{T}^2$. Denote the coordinates by t, θ_1 and θ_2 respectively. The 1-form $\alpha = \cos t\, d\theta_1 + \sin t\, d\theta_2$ is contact. [Check this.]

Lemma I.19. *Suppose α is a contact form on a manifold B. Then for any positive function f on B the 1-form $f\alpha$ is also contact.*

Proof. Note first that since f is nowhere zero, $\ker f\alpha = \ker \alpha$. Thus to show that $f\alpha$ is contact, it is enough to check that $d(f\alpha)|_\xi$ is nondegenerate, where $\xi = \ker \alpha = \ker f\alpha$. Now $d(f\alpha) = df \wedge \alpha + f\, d\alpha$ and $\alpha|_\xi = 0$. Therefore $d(f\alpha)|_\xi = f\, d\alpha|_\xi$. But f is nowhere zero and $d\alpha|_\xi$ is nondegenerate by assumption. Thus $d(f\alpha)|_\xi$ is nondegenerate. $\square$

Definition I.20. We define the **conformal class** of a 1-form α on a manifold B to be the set $[\alpha] = \{e^h \alpha \mid h \in C^\infty(B)\}$, that is, the set of all 1-forms obtained from α by multiplying it by a positive function.

Thus if a 1-form α on a manifold B is contact, then its conformal class consists of contact forms all defining the same subbundle ξ of the tangent bundle of B.

Definition I.21. A (co-orientable) **contact structure** ξ on a manifold B is a subbundle of the tangent bundle TB of the form $\xi = \ker \alpha$ for some contact form α.

A **co-orientation** of a contact structure ξ is a choice of a conformal class of contact forms defining the contact structure.

Remark I.22. More generally a **contact structure** on a manifold B is a subbundle ξ of the tangent bundle TB such that for every point $x \in B$ there is a contact 1-form α defined in a neighborhood of x with $\ker \alpha = \xi$. There exist contact structures which are not co-orientable. For such structures ξ a one-form α with $\ker \alpha = \xi$ exists only **locally**. IN THESE NOTES WE WILL ONLY DEAL WITH CO-ORIENTABLE CONTACT STRUCTURES .

Exercise I.23. Let β be a nowhere zero 1-form on a manifold B and let $\eta = \ker \beta$. Let $\eta^\circ \to B$ denote the annihilator of η in T^*B: the fiber of η° at $b \in B$ is the vector space

$$\eta_b^\circ = \{p \in T_b^*B \mid p|_{\eta_b} = 0\}.$$

Show that β is a nowhere zero section of real line bundle $\eta^\circ \to B$. Show that any other nowhere zero section β' of $\eta^\circ \to B$ is of the form $\beta' = f\beta$ for some nowhere zero function f on B.

Conclude that a contact structure ξ is co-orientable if and only if the punctured real line bundle $\xi^\circ \smallsetminus 0$ has two components (0 of course denotes the zero section). Show that a choice of co-orientation of ξ is the same as a choice of a component ξ_+° of the punctured bundle $\xi^\circ \smallsetminus 0$.

Definition I.24. Let $(B_1, \xi_1 = \ker \alpha_1)$ and $(B_2, \xi_2 = \ker \alpha_2)$ be two co-oriented contact manifolds. A diffeomorphism $\varphi : B_1 \to B_2$ is a **contactomorphism** if the differential $d\varphi$ maps ξ_1 to ξ_2 preserving the co-orientations. That is, $\varphi^*\alpha_2 = f\alpha_1$ for some positive function f.[4]

Definition I.25. An action of a Lie group G on a manifold B **preserves a contact structure** ξ and its co-orientation if for every element $a \in G$ the corresponding diffeomorphism $a_B : B \to B$ is a contactomorphism. We will also say that the action of G on (B, ξ) is a **contact action**.

Definition I.26. Let α be a contact form on a manifold B. The **Reeb** vector field Y_α of α is the unique vector field satisfying $\iota(Y_\alpha)d\alpha = 0$ and $\alpha(Y_\alpha) = 1$.

Exercise I.27. Why does the definition of the Reeb vector field makes sense?

Remark I.28. The Reeb vector field depends strongly on the contact form. And it is not just its magnitude: if α is a contact form and Y_α is its Reeb vector field, then there is no reason for $\iota(Y_\alpha)d(f\alpha) = 0$ where f is a nowhere zero function.

Exercise I.29. Compute the Reeb vector field of the contact form $\alpha = dz + x\,dy$ on $\mathbb{R}^3$.

Exercise I.30. Compute the Reeb vector field of the contact form α of Example I.18: $\alpha = \cos t\,d\theta_1 + \sin t\,d\theta_2$ on $B = \mathbb{R} \times \mathbb{T}^2$.

Symplectic cones and contact manifolds are intimately related:

Theorem I.31. *Suppose a compact connected Lie group G acts effectively on a symplectic cone (M, ω, X) preserving the symplectic form ω and the expanding vector field X. Then G induces an effective action on the base B of the cone making the projection $\varpi : M \to B$ G-equivariant. Moreover, the base B has a natural co-oriented contact structure ξ, and the induced action of G on B preserves a contact form α defining ξ. In particular the action of G on (B, ξ) is contact.*

---

[4] Since $d\varphi(\xi_1) = \xi_2$, the lift $\tilde{\varphi} : T^*B_1 \to T^*B_2$ of φ maps ξ_1° to ξ_2°. "$d\varphi$ preserves the co-orientation" means that $\tilde{\varphi}$ maps $(\xi_1)_+^\circ$ to $(\xi_2)_+^\circ$ (cf. Exercise I.23).

We start the proof of Theorem I.31 with an observation that the action of G on M descends to an effective action of G on the base B making the projection $\varpi : M \to B$ G-equivariant. Next we prove:

Proposition I.32. *Any principal $\mathbb{R}$-bundle $\mathbb{R} \to M \xrightarrow{\varpi} B$ is trivial.*

Proof. The proposition is true because the real line is contractible. Here is an elementary argument. Note first that if $s : B \to M$ is a (local) section of $\varpi : M \to B$ and $f \in C^\infty(B)$ is a function, then $s - f$ makes sense; it is again a (local) section of $\varpi : M \to B$. To prove that a principal bundle is trivial it is enough to construct a global section. To this end choose an open cover $\{U_\alpha\}$ of B such that for each U_α there is a section $s_\alpha : U_\alpha \to M$. Choose a partition of unity τ_α subordinate to the cover $\{U_\alpha\}$. Two sections of a principal $\mathbb{R}$-bundle differ by real-valued function. Thus by abuse of notation on an intersection $U_\alpha \cap U_\beta$, $s_\alpha - s_\beta$ is a real-valued function. Now define for each index α

$$s'_\beta = s_\beta - \sum_{\alpha \neq \beta} \tau_\alpha(s_\beta - s_\alpha).$$

Then on an intersection $U_\alpha \cap U_\beta$

$$
\begin{aligned}
s'_\beta - s'_\gamma &= \left(s_\beta - \sum_{\alpha \neq \beta} \tau_\alpha(s_\beta - s_\alpha) \right) - \left(s_\gamma - \sum_{\alpha \neq \gamma} \tau_\alpha(s_\gamma - s_\alpha) \right) \\
&= s_\beta - s_\gamma - \left(\sum_{\alpha \neq \beta, \gamma} \tau_\alpha(s_\beta - s_\gamma) \right) + \tau_\beta(s_\gamma - s_\beta) - \tau_\gamma(s_\beta - s_\gamma) \\
&= s_\beta - s_\gamma - \left(\sum_\alpha \tau_\alpha \right)(s_\beta - s_\gamma) = 0.
\end{aligned}
$$

Therefore the collection of local sections $\{s'_\alpha\}$ defines a global section of $\varpi : M \to B$. Consequently the bundle is trivial. $\square$

Thus any symplectic cone is of the form $B \times \mathbb{R}$ where $B = M/\mathbb{R}$ is an odd-dimensional manifold.

Lemma I.33. *Let (M, ω, X) be a symplectic cone, let B be its base and let $\varpi : M \to B$ denote the projection. Pick a trivialization $\varphi : B \times \mathbb{R} \to M$. Then $\varphi^* \omega = d(e^t \alpha)$ where t is a coordinate on $\mathbb{R}$ and α is a contact form on B. Conversely, if α is contact form on B then $(B \times \mathbb{R}, d(e^t \alpha), \frac{\partial}{\partial t})$ is a symplectic cone.*

Proof. By Proposition I.32 the principal $\mathbb{R}$ bundle $\varpi : M \to B$ is trivial. Choose a trivialization $M \simeq B \times \mathbb{R}$. Under this identification the vector field X becomes $\frac{\partial}{\partial t}$.

Since $d\omega = 0$ and $L_X \omega = \omega$, $d\iota(X)\omega = \omega$ (c.f. proof of Proposition I.13). Let $\beta = \iota(X)\omega$. Then $\iota(X)\beta = 0$ and $L_X \beta = d\iota(X)\beta + \iota(X)d\beta = 0 + \iota(X)\omega = \beta$.

Hence for any point $(b, t) \in B \times \mathbb{R}$

$$\beta_{(b,t)} = e^t \beta_{(b,0)}.$$

Since $\iota(X(b, 0))\beta_{(b,0)} = 0$ it follows that $\beta_{(b,0)} = \alpha_b$ for a 1-form α on B. It remains to show that α is contact. For this it suffices to show that $\alpha \wedge (d\alpha)^d$ is nowhere zero, where $d = \frac{1}{2} \dim M - 1$. Now $\omega = d(e^t \alpha)$ is symplectic. Hence ω^{d+1} is nowhere vanishing. Now $\omega^{d+1} = (e^t(dt \wedge \alpha + d\alpha))^{d+1} = e^{td}dt \wedge \alpha \wedge (d\alpha)^d$. Hence $\alpha \wedge (d\alpha)^d$ is nowhere vanishing.

Conversely suppose α is a contact 1-form on B. Let $\omega = d(e^t \alpha)$ and let $X = \frac{\partial}{\partial t}$. Then $L_X \omega = d(\iota(\frac{\partial}{\partial t})d(e^t \alpha)) = d(\iota(\frac{\partial}{\partial t})(e^t dt \wedge \alpha + e^t d\alpha)) = d(e^t \alpha + 0) = \omega$. It remains to check that ω is nondegenerate. For any $(b, t) \in B \times \mathbb{R}$, the tangent space $T_{(b,t)}(B \times \mathbb{R})$ decomposes as $T_{(b,t)}(B \times \mathbb{R}) = \ker \alpha_b \oplus \mathbb{R}Y_\alpha(b) \oplus \mathbb{R}$ where Y_α is the Reeb vector field of α (cf. Definition I.26). Since α is contact $d\alpha_b|_{\ker \alpha_b}$ is nondegenerate. The restriction $dt \wedge \alpha_b$ to $\mathbb{R}Y_\alpha(b) \oplus \mathbb{R}$ is nondegenerate as well. Hence $\omega = e^t(dt \wedge \alpha + d\alpha)$ is nondegenerate. This proves that $(B \times \mathbb{R}, d(e^t \alpha), \frac{\partial}{\partial t})$ is a symplectic cone. $\qquad \square$

Exercise I.34. Let (M, ω, X) be a symplectic cone, let B be its base and let $\varpi : M \to B$ denote the $\mathbb{R}$-orbit map. Pick a global section $s : B \to M$ of $\varpi : M \to B$ and let $\alpha = s^*(\iota(X)\omega)$. Show that α is a contact form on B. Show that it is the same contact form that the proof of Lemma I.33 would produce from the trivialization $\varphi : B \times \mathbb{R} \to M$, $\varphi(b, t) = \rho_t(s(b))$. Here $\rho_t : M \to M$ denotes the action of $\mathbb{R}$.

Remark I.35. Different choices of trivializations φ of $\varpi : M \to B$ give rise to different contact forms on B. However, they all define the same contact structure ξ on the base B. Intrinsically ξ can be defined as follows: for a point $b \in B$,

$$\xi_b = d\varpi_m(\ker(\iota(X)\omega)_m) \quad \text{for any } m \in \varpi^{-1}(b) \tag{I.1}$$

It is not hard to check that ξ is well-defined. First note that $\mathbb{R}$ acts transitively on the fiber $\varpi^{-1}(b)$. Second observe that for any $\lambda \in \mathbb{R}$ we have $\rho_\lambda^*(\iota(X)\omega) = e^\lambda(\iota(X)\omega)$, and hence $d\rho_\lambda(\ker(\iota(X)\omega)_m) = (\ker(\iota(X)\omega)_{\rho_\lambda(m)}$. Here again $\rho_t : M \to M$ denotes the action of $\mathbb{R}$.

It follows that the action of G on B induced by an action of G on the symplectic cone $\varpi : M \to B$ preserves the contact structure ξ defined by (I.1). Since G is connected and since the identity map preserves the co-orientation of ξ, all the other elements of G also preserve the co-orientation.

It remains to show that there is a G-invariant 1-form α with $\ker \alpha = \xi$. By Lemma I.33 a choice of a trivialization of $\varpi : M \to B$ gives us a 1-form α on B with $\ker \alpha = \xi$, but this form need not be G-invariant. It is only G-invariant if the trivialization is G-equivariant. Therefore we proceed as follows.

Lemma I.36. *Suppose a compact Lie group G acts on a manifold B preserving a (co-oriented) contact structure $\xi = \ker \alpha$ for some 1-form α. Then there exists a G-invariant 1-form $\tilde{\alpha}$ with $\ker \tilde{\alpha} = \xi$.*

Proof. For every $a \in G$ the corresponding diffeomorphism $a_B : B \to B$ is a contactomorphism. Hence $(a_B)^*\alpha = f_a \alpha$ for some positive function f_a depending smoothly on a. Define a new contact form $\tilde{\alpha}$ to be the average of α over the action of G:

$$\tilde{\alpha}_b = \int_G ((a_B)^*\alpha)_b \, da = \int_G (f_a(b)\alpha_b) \, da = \left(\int_G f_a(b) \, da \right) \alpha_b$$

for all $b \in B$. Here da is a bi-invariant measure on G normalized so that $\int_G da = 1$. Since $f_a > 0$ for all $a \in G$, the integral $\left(\int_G f_a(b) \, da \right)$ is positive and $\tilde{\alpha}$ is indeed nowhere zero.   □

    From now on we will always assume that whenever a group actions preserves a contact structure it also preserves a contact form defining this structure.

    This concludes the proof of Theorem I.31. It follows from the Theorem that if an n-torus G acts effectively on the punctured cotangent bundle $M = T^*\mathbb{T}^n \setminus 0$ preserving the symplectic form and commuting with dilations then it acts on the quotient $B = M/\mathbb{R} \simeq S^*\mathbb{T}^n$ preserving the corresponding contact structure. Note that $2 \dim G = \dim B + 1$.

Definition I.37. An effective action of a torus G on a manifold B preserving a contact structure ξ is **completely integrable** if $2 \dim G = \dim B + 1$.

    A **contact toric G-manifold** is a co-oriented contact manifold with a completely integrable action of a torus G.

    We are now in position to reduce Theorem I.4 to a statement about contact toric manifolds. Consider again the action of $G = \mathbb{T}^n$ on $M = T^*\mathbb{T}^n \setminus 0$ preserving the symplectic form and commuting with dilations. As was remarked previously M is a symplectic cone over $B = S^*\mathbb{T}^n = \mathbb{T}^n \times S^{n-1}$. By Theorem I.31 the action of G on M induces an effective action on B. Moreover B has a G-invariant contact structure ξ making (B,ξ) into a compact connected contact toric G-manifold. Clearly if the action of G on B is free then the original action of G on M was free as well. Therefore a proof of Theorem I.4 reduces to

Theorem I.38. *Let $(B, \xi = \ker \alpha)$ be a compact connected contact toric G-manifold and suppose the action of G is not free. Then B is not homotopy equivalent to $\mathbb{T}^n \times S^{n-1}$, $n = \dim G$.*

    The rest of the notes will be occupied with a proof of Theorem I.38. The proof uses heavily contact moment maps which are discussed in the next chapter. We end this section with a partial converse to Theorem I.31.

    We have seen that given a contact form α on a manifold B the form $d(e^t\alpha)$ on $B \times \mathbb{R}$ is symplectic. The pair $(B \times \mathbb{R}, d(e^t\alpha))$ is called the **symplectization**[5] of (B, α). It is clearly a symplectic cone.

    Different contact forms on B give rise to different symplectic forms on $B \times \mathbb{R}$. However there is a symplectic cone that depends only on the contact structure

---

[5]Sometimes $(B \times \mathbb{R}, d(e^t\alpha))$ is called the **symplectification** of (B, α).

$\xi = \ker \alpha$ (and its co-orientation) and not on a particular choice of a contact form: Let ξ_+° denote the component of $\xi^\circ \setminus 0$ giving ξ its co-orientation (cf. Exercise I.23). It is not hard to check that ξ_+° is a symplectic submanifold of the cotangent bundle T^*B with its standard symplectic structure. The action of $\mathbb{R}$ on $T^*B \setminus 0$ by dilations preserves ξ_+° and makes it into a symplectic cone. A section β of $\xi_+^\circ \to B$ is a contact form β on B with $\xi = \ker \beta$. The form β defines a trivialization $\varphi_\beta : B \times \mathbb{R} \to \xi_+^\circ$, $\varphi_\beta(b, t) = e^t \beta_b$. The map φ_β pulls back the symplectic form on ξ_+° to $d(e^t \beta)$.

Finally given a symplectic cone (M, ω, X) with the base B, the orbit map $\varpi : M \to B$ and the induced contact structure ξ on B, there is an $\mathbb{R}$ equivariant symplectomorphism $\varphi : M \to \xi_+^\circ$ defined as follows: for a point $m \in M$ let $\varphi(m)$ be the covector in $T^*_{\varpi(m)}B$ such that

$$\varphi(m)(v) = (\iota(X)\omega)_m(ds_b(v))$$

for all $v \in T_{\varpi(m)}B$ and a section s of $\varpi : M \to B$.

The discussion above can be summarized as:

Lemma I.39. *Let $(B, \xi = \ker \alpha)$ be a contact manifold, let ξ_+° be a component of $\xi^\circ \setminus 0$, the annihilator of ξ in T^*B minus the zero section. The principal $\mathbb{R}$ bundle $\xi_+^\circ \to B$ is a symplectic cone.*

If (M, ω, X) is a symplectic cone with the base B and ξ is the induced contact structure on B, then ξ_+° is isomorphic to (M, ω, X) as a symplectic cone.

Chapter II

Contact group actions and contact moment maps

Moment maps exist in the category of contact group actions. In fact moment maps exist for all contact actions. This is because a contact form defines a bijection between contact vector fields and smooth functions.

Definition II.1. A vector field X on a contact manifold $(B, \xi = \ker \alpha)$ is **contact** if its flow φ_t consists of contactomorphisms. In particular $d\varphi_t(\xi) \subset \xi$. Hence for any section v of the bundle $\xi \to B$, the Lie bracket $[X, v]$ is again a section of $\xi \to B$.

Thus for a contact action of a Lie group G on (B, ξ) the vector fields induced by elements of the Lie algebra $\mathfrak{g}$ of G are contact.

Exercise II.2. Prove that a Reeb vector field is contact. More generally prove that a vector field X on a contact manifold $(B, \xi = \ker \alpha)$ is contact if and only if $L_X \alpha = h\alpha$ for some function h (h can have zeros).

A choice of a contact form on a contact manifold (B, ξ) identifies contact vector fields with smooth functions.

Proposition II.3. *Let $(B, \xi = \ker \alpha)$ be a contact manifold. The linear map from contact vector fields to smooth functions given by $X \mapsto f^X := \alpha(X)$ is one-to-one and onto.*

Proof. Note that the Reeb vector field Y_α corresponds to the function 1. For any vector field X on B the vector field $X - \alpha(X)Y_\alpha$ is in the kernel of α, which is the contact distribution ξ. Since $d\alpha|_\xi$ is non-degenerate, $X - \alpha(X)Y_\alpha$ is uniquely determined by $\iota(X - \alpha(X)Y_\alpha)(d\alpha|_\xi)$.

For any section v of $\xi \to B$ and any vector field X we have

$$0 = L_X 0 = L_X(\alpha(v)) = (L_X \alpha)(v) + \alpha([X, v]).$$

Assume now that X is contact. Then $\alpha([X, v]) = 0$. Therefore, by Cartan's formula

$$0 = (d\iota(X)\alpha + \iota(X)d\alpha)(v).$$

Hence for any section v of ξ, $d\alpha(X, v) = -d(\alpha(X))(v) = -df^X(v)$. And, of course, $d\alpha(X, v) = d\alpha(X - f^X Y_\alpha, v)$ for all v. We conclude that

$$\iota(X - \alpha(X)Y_\alpha)(d\alpha|_\xi) = -df^X|_\xi \qquad\qquad (\text{II}.1)$$

for any contact vector field X. Thus if X is contact, its component in the direction of the Reeb vector field is $f^X Y_\alpha$ and its component in the direction of the contact distribution is uniquely determined by (II.1).

Conversely, given a function f on B there is a unique section X_f' of ξ such that

$$\iota(X_f')d\alpha|_\xi = -df|_\xi. \qquad\qquad (\text{II}.2)$$

The vector field $X_f := X_f' + fY_\alpha$ is a contact vector field with $\alpha(X_f) = f$. $\square$

Exercise II.4. Given a function f on a manifold B with contact form α check that the vector field $X_f := X_f' + fY_\alpha$, where X_f' is defined by (II.2), is contact. That is, show that $L_{X_f}\alpha = h\alpha$ for some function h (cf. Exercise II.2).

Suppose now that a Lie group G acts on a manifold B preserving a contact 1-form α. Then for any vector A in the Lie algebra $\mathfrak{g}$ of G, the induced vector field A_B satisfies[1] $L_{A_B}\alpha = 0$ and, in particular, is contact. Since the contact form α defines a 1-1 correspondence between contact vector fields and functions, it makes sense to define the α-**moment map** $\Psi_\alpha : B \to \mathfrak{g}^*$ by

$$\langle \Psi_\alpha(b), A \rangle = \alpha_b(A_B(b)) \qquad\qquad (\text{II}.3)$$

for all $b \in B$ and all $A \in \mathfrak{g}$. Note that by Exercise I.14 the α-moment map Ψ_α is G-equivariant.

The α-moment map, as the name suggests, depends rather strongly on α: if f is any positive G-invariant function on B, then $f\alpha$ is another G-invariant contact form and clearly

$$\Psi_{f\alpha} = f\Psi_\alpha.$$

In particular, unlike in the symplectic case, the image of a moment map is **not** an invariant of the action and of the contact structure. However the **moment cone** $C(\Psi_\alpha)$ defined by

$$C(\Psi_\alpha) = \{t\eta \in \mathfrak{g}^* \mid \eta \in \Psi_\alpha(B), \, t \in [0, \infty)\}$$

is an invariant of the action of G on $(B, \xi = \ker\alpha)$.

---

[1] Recall that the vector field A_B is defined by $A_B(b) = \frac{d}{dt}\big|_{t=0} \exp(tA) \cdot b$.

Remark II.5. Suppose a Lie group G acts on a manifold B preserving a contact form α. The action of G lifts to an action on the cotangent bundle T^*B which preserves the annihilator ξ° of the contact structure $\xi = \ker \alpha$. Moreover, the action preserves the component ξ°_+ of $\xi \smallsetminus 0$ which contains the image of $\alpha : B \to T^*B$.

The action of G on T^*B is Hamiltonian with a natural moment map $\Phi : T^*B \to \mathfrak{g}^*$ given by

$$\langle \Phi(b,p), A \rangle = \langle p, A_B(b) \rangle$$

for all $b \in B$, $p \in T^*_b B$, $A \in \mathfrak{g}$. Since the submanifold ξ°_+ is a G-invariant symplectic submanifold of T^*B the restriction $\Psi := \Phi|_{\xi^\circ_+}$ is a moment map for the action of G on ξ°_+. It is not hard to check that

$$\Psi_\alpha = \Psi \circ \alpha.$$

For this and other reasons it makes sense to think of $\Psi : \xi^\circ_+ \to \mathfrak{g}^*$ as the moment map for the action of G on (B, ξ).[2] Note also that $C(\Psi_\alpha) = \Psi(\xi^\circ_+) \cup \{0\}$. We will thus denote the moment cone $C(\Psi_\alpha)$ by $C(\Psi)$.

Consequently and by analogy with symplectic toric manifolds we will think of **contact toric G-manifolds** as triples $(B, \xi = \ker \alpha, \Psi : \xi^\circ_+ \to \mathfrak{g}^*)$.

---

[2]Note that since Φ is G-equivariant and $\alpha : B \to T^*B$ is G-equivariant, Ψ_α is G-equivariant as well. This gives us an alternative proof that the α-moment map is equivariant.

Chapter III

Proof of Theorem I.38

The rest of the lecture notes will be devoted to a proof of Theorem I.38. Right from the beginning the proof will bifurcate into two cases: the contact manifold B is 3-dimensional and $\dim B > 3$. If $\dim B = 3$ we will argue directly using slices that the orbit space B/G is homeomorphic to a closed interval $[0,1]$ and then use this to compute the integral cohomology of B. This will show that B cannot be homeomorphic to $S^*\mathbb{T}^2 = \mathbb{T}^3$.

We will then consider the case where $\dim B > 3$. In this case we have a connectedness and convexity theorem of Banyaga and Molino (see [BM1, BM2]; for a different proof see [L2]):

Theorem III.1. *Let $(B, \xi = \ker \alpha, \Psi : \xi_+^\circ \to \mathfrak{g}^*)$ be a compact connected contact toric G-manifold. Suppose $\dim B > 3$. Then the fibers of the moment map Ψ are G-orbits (and in particular are connected) and the moment cone $C(\Psi)$ is a convex polyhedral cone in $\mathfrak{g}^*$. Moreover $C(\Psi) \neq \mathfrak{g}^*$ iff the action of G is not free.*

Our proof will then bifurcate again. We will consider separately the case where the moment cone contains a linear subspace of dimension k, $0 < k < \dim G$ and where no such subspace exists (i.e., the moment cone is **proper**).

In the first case we will use a uniqueness theorem of Boucetta and Molino [BoM][1] for symplectic toric manifolds to argue that the symplectization $B \times \mathbb{R}$ of B is diffeomorphic to the manifold $N = \mathbb{T}^k \times (\mathbb{R}^k \times \mathbb{C}^l \smallsetminus \{(0,0)\})$ where $2l + 2k = \dim B + 1$ and $k, l > 0$. It is easy to see that N cannot be homotopy equivalent $T^*\mathbb{T}^n \smallsetminus 0$, $n = k + l$.

In the latter case we will argue following Boyer and Galicki [BG] that there is a locally free[2] S^1 action on B such that the quotient B/S^1 is a (compact connected) symplectic toric orbifold. Since the action of S^1 is locally free there is

---

[1] The result was rediscovered a few years later by Lerman, Tolman and Woodward (Lemma 7.2 and Proposition 7.3 in [LT]).

[2] An action of a Lie group G on a manifold Z is locally free if all the isotropy groups are zero dimensional.

a long exact sequence of rational cohomology groups (the Gysin sequence) tying together the cohomology of B and of B/S^1. On the other hand Morse theory on the orbifold B/S^1 shows that all odd-dimensional rational cohomology of B/S^1 vanishes. Together these two facts will imply that $\dim_{\mathbb{Q}} H^1(B, \mathbb{Q}) \leq 1$. Thus B cannot be $S^*\mathbb{T}^n = \mathbb{T}^n \times S^{n-1}$, $n = \frac{1}{2}(\dim B + 1) > 2$. This completes the preview of our proof of Theorem I.38.

III.1   Homogeneous vector bundles and slices

Suppose that G is a Lie group, $H \subset G$ a closed subgroup and suppose we have a representation of H on a vector space W. Then H acts on the product $G \times W$ by $h \cdot (g, w) = (gh^{-1}, g \cdot w)$ for $h \in H$, $(g, w) \in G \times W$. The quotient $G \times_H W :=$ $(G \times W)/H$ is a vector bundle over G/H with typical fiber W. We denote the image of $(g, w) \in G \times W$ in $G \times_H W$ by $[g, w]$. Note that the action of G on $G \times W$ given by $a \cdot (g, w) = (ag, w)$ commutes with the action of H and hence descends to an action G on $G \times_H W$: $a \cdot [g, w] = [ag, w]$. The projection $G \times_H W \to G/H$ is G-equivariant and the action of G on the base G/H is transitive. This makes $G \times_H W \to G/H$ into a homogeneous vector bundle.

Conversely if $\pi : E \to G/H$ is a vector bundle with an action of G by vector bundle maps making π equivariant, then E is isomorphic to $G \times_H W$ where W is the fiber of E above the identity coset $1H$.[3] Indeed the map

$$G \times W \to E, \quad (g, w) \mapsto g \cdot w$$

is onto and is constant along the orbits of H. It descends to a vector bundle isomorphism

$$G \times_H W \to E, \quad [g, w] \mapsto g \cdot w.$$

Next suppose a compact Lie group G acts on a manifold M. Consider an orbit $G \cdot x \subset M$. The group G acts on the normal bundle of the orbit $\nu(G \cdot x) = TM|_{G \cdot x}/T(G \cdot x)$ making the projection $\pi : \nu(G \cdot x) \to (G \cdot x)$ equivariant. Thus $\nu(G \cdot x) = G \times_H W$ where $W = T_x M/T_x(G \cdot x)$.

If we choose a G-invariant Riemannian metric on M [4] we can identify $\nu(G \cdot x)$ with the perpendicular of $T(G \cdot x)$ in $TM|_{G \cdot x}$. Furthermore, the Riemannian exponential map $\exp : \nu(G \cdot x) \to M$ is G-equivariant. Hence, by the Tubular Neighborhood theorem we get:

Lemma III.2. *Let G be a compact connected Lie group acting on a manifold M and let $x \in M$ be a point. A neighborhood of $G \cdot x$ in M is G-equivariantly diffeomorphic to a neighborhood of the zero section of the homogeneous vector bundle $G \times_{G_x} W$ where G_x denotes the isotropy group of x and $W = T_x M/T_x(G \cdot x)$.*

---

[3] Why is there a representation of H on W?

[4] Choose any metric on M and then average it over the group G (cf. proof of Lemma I.36).

Definition III.3. Let G be a Lie group acting on a manifold M; let $x \in M$ be a point. Denote the isotropy group of x by G_x. An embedded submanifold $S \subset M$ is a **slice through** x **for the action of** G **on** M if $x \in S$, S is G_x-invariant and if the map $G \times S \to M$ given by $(g, s) \mapsto g \cdot s$ descends to an open embedding $G \times_{G_x} S \to M$, $[g, s] \mapsto g \cdot s$.

Thus by Lemma III.2 slices exist for actions of compact Lie groups: we may choose as a slice at x the image of a small G_x-invariant neighborhood of 0 in $W = T_x M / T_x(G \cdot x)$ under the exponential map.

Here is a typical application of the existence of slices. Locally near x the quotient M/G is homeomorphic to the quotient $(G \times_{G_x} W)/G = W/G_x$. Hence quotients of manifolds by actions of compact Lie groups are modeled on quotients of vector spaces by linear actions of compact Lie groups. The linear action of G_x on $W = T_x M / T_x(G \cdot x)$ is called the **slice representation** at x.

Here is another application of the above construction:

Lemma III.4. *Suppose G is a compact abelian group acting effectively on a connected manifold M. Then every slice representation is **faithful**, i.e., no slice representation has a kernel.*

Proof. Suppose the slice representation of $H = G_x$ on $W = T_x M / T_x(G \cdot x)$ is not faithful at a point $x \in M$. By Lemma III.2 it is no loss of generality to assume that a neighborhood of $G \cdot x$ in M is the homogeneous vector bundle $G \times_H W \to G/H = G \cdot x$ and that the point x is $[1, 0] \in G \times_H W$. If there is an element $a \in H$ such that $a \neq 1$ and yet $a \cdot w = w$ for all $w \in W$, then $a \cdot [g, w] = [ag, w] = [ga, w] = [g, a \cdot w] = [g, w]$ for all $[g, w] \in G \times_H W$ ($ag = ga$ since G is abelian). Thus $a \in H$ fixes an open neighborhood of x. Since the set fixed by a is closed and since M is connected it follows that a fixes all of M. This contradicts the assumption that the action of G on M is effective. □

Remark III.5. The compactness of the Lie group G is not necessary for the existence of slices. According to Palais [P] it is only necessary that its action on a manifold M be proper (see Definition I.9 above).

Thus if an action of a Lie group G on a manifold M is free and proper, then the existence of slices tell us that a neighborhood of every orbit is equivariantly diffeomorphic to a product of G with some manifold S. It is not hard to deduce from this that the orbit space M/G is a manifold and that the orbit map $M \to M/G$ makes M into a principal G-bundle.

We now recall a few properties of tori, which for us are compact abelian Lie groups. If G is a torus, then the (Lie group) exponential map $\exp : \mathfrak{g} \to G$ is a covering map. The kernel $\mathbb{Z}_G$ of $\exp$ is called the **integral lattice**. Clearly $G = \mathfrak{g}/\mathbb{Z}_G$. The group $\mathbb{Z}_G$ is isomorphic to the fundamental group of G. Also it has the property that for any $X \in \mathbb{Z}_G$ the corresponding 1-parameter subgroup $\{\exp tX \mid t \in \mathbb{R}\}$ is a circle. The dual lattice $\mathbb{Z}_G^* = \mathrm{Hom}_{\mathbb{Z}}(\mathbb{Z}_G, \mathbb{Z}) \cong \{\ell \in \mathfrak{g}^* \mid \ell(\mathbb{Z}_G) \subset \mathbb{Z}\}$ is the **weight lattice**. It parameterizes 1-dimension complex representations of G, or, equivalently, group homomorphisms (**characters**) $\chi : G \to S^1$: Given $\nu \in \mathbb{Z}_G^*$ the corresponding character $\chi_\nu : G = \mathfrak{g}/\mathbb{Z}_G \to S^1$ is defined by $\chi_\nu(\exp X) = e^{2\pi i \nu(X)}$ for all $X \in \mathfrak{g}$. Given a character $\chi : G \to S^1$, its differential $d\chi_1$ at 1 is a weight.

Recall that a complex representation of a compact abelian group is a direct sum of one-dimensional complex representations. Thus a complex representation of a torus is completely characterized by a finite set of weights. The same is true for *symplectic* representations of tori — one defines weights with respect to some complex structure compatible with the symplectic form. The weights do not depend on the choice of the complex structure; they only depend on the symplectic form.

The Lemma below is a key representation-theoretic fact in the classification of symplectic and contact toric manifolds.

Lemma III.6. *Suppose $\rho : H \to Sp(V, \omega)$ is a symplectic representation of a compact abelian Lie group H on a symplectic vector space (V, ω). Suppose that ρ is faithful, i.e., $\ker \rho = \{1\}$.*

Then $\dim H \leq \frac{1}{2} \dim V$. If $H = \frac{1}{2} \dim V$, then H is connected. Moreover, the set of weights for the representation of H on (V, ω) is a basis of the weight lattice $\mathbb{Z}_H^*$ of H.

Proof. Since H is compact there exists an H-invariant complex structure J on V compatible with ω (i.e., $\omega(J\cdot, J\cdot) = \omega(\cdot, \cdot)$ and for any $v \neq 0$ we have $\omega(Jv, v) > 0$).

The choice of J identifies (V, ω) with $\mathbb{C}^n$ with the standard symplectic structure $\sqrt{-1} \sum dz_j \wedge d\bar{z}_j$, $n = \frac{1}{2} \dim_{\mathbb{R}} V$. This, in turn, gives us a representation of H on $\mathbb{C}^n$ by unitary matrices. Thus we may assume that ρ is an injective group homomorphism $\rho : H \to U(n)$.

The connected component H° of H is a torus. Therefore $\rho(H^\circ)$ is contained in a maximal torus of $U(n)$ which is the n-torus. Since ρ has no kernel we have $\dim H = \dim H^\circ = \dim \rho(H^\circ) \leq n = \frac{1}{2} \dim V$.

Now suppose $\dim H = n$. Since all maximal tori in $U(n)$ are conjugate, we may assume that $\rho(H^\circ)$ is the standard maximal torus, that is $\rho(H^\circ)$ is the set of all diagonal unitary matrices. Since the only unitary matrices which commute with all the diagonal matrices are the diagonal matrices, we see that we must have $\rho(H) = \rho(H^\circ)$. Consequently since ρ is faithful, $H = H^\circ$, i.e., H is a torus.

Finally the set of weights of the maximal torus $\mathbb{T}^n$ in $U(n)$ for its representation on $\mathbb{C}^n$ is a lattice basis of weight lattice $\mathbb{Z}_{\mathbb{T}^n}^*$. Hence the set of weights for the representation of H on (V, ω) is a basis of the weight lattice $\mathbb{Z}_H^*$ of H. □

Lemma III.7. *Let $(B, \xi = \ker \alpha)$ be a contact toric G-manifold. Then*

1. *No G-orbit is tangent to the contact structure ξ. In particular there are no fixed points. Hence the α-moment map Ψ_α does not vanish at any point for any G-invariant contact form α. Equivalently $\Psi(\xi_+^\circ)$ does not contain the origin in $\mathfrak{g}^*$.*

2. *All isotropy groups are connected.*

Proof. Let b be a point in B and let H denote its isotropy group. The group H acts on the tangent space $T_b B$. Since the contact form α is G-invariant, its kernel at b, the hyperplane ξ_b, is an H-invariant subspace of $T_b B$. Since the Reeb vector field Y_α of α is unique, the vector $Y_\alpha(b)$ is fixed by H. Thus we have an H-equivariant splitting

$$T_b B = \mathbb{R} Y_\alpha(b) \oplus \xi_b.$$

Let $V = T_b(G \cdot b) \cap \xi_b$. It is an H-invariant subspace of ξ_b. Note that since $\dim \xi_b = \dim B - 1$ we either have $V = T_b(G \cdot b)$ or $\dim V = \dim(G \cdot b) - 1$. We will argue that the former case cannot occur. But first we argue that V is an isotropic subspace of the symplectic vector space $(\xi_b, \omega = d\alpha_b|_{\xi_b})$.

Let $x, z \in V$ be two vectors. There exist vectors $X, Z \in \mathfrak{g}$ such that $X_B(b) = x$ and $Z_B(b) = z$. Then $\omega(x, z) = d\alpha_b(X_B(b), Z_B(b))$. Since G is abelian $[Z_B, X_B] = 0$. The function $\alpha(X_B)$ is G-invariant, hence $Z_B(\alpha(X_B)) = 0$. Therefore

$$\omega(x, y) = d\alpha(X_B, Z_B) = X_B(\alpha(Z_B)) - Z_B(\alpha(X_B)) - d\alpha([X_B, Z_B]) = 0 - 0 + 0.$$

This proves that V is isotropic.

Since H is compact, there is an H-invariant complex structure J on ξ_b compatible with ω. Since V is isotropic, $V \cap JV = 0$ and $V + JV = V \oplus JV$ is a symplectic subspace of (ξ_b, ω). It is H-invariant. In fact, since G is abelian, the action of H on $T_b(G \cdot b)$ is trivial. Hence the action of H on $V \oplus JV$ is trivial as well. Let W denote the symplectic perpendicular to $V \oplus JV$. We get a symplectic representation of H on W.

We now argue that if $T_b(G \cdot b) \subset \xi_b$ then $\dim W$ is less than $2 \dim H$ and that the representation of H on W must be faithful. This by Lemma III.6 would give us a contradiction. We would then conclude that $\dim W = 2 \dim H$, which by Lemma III.6 implies that H is connected.

Suppose now that $T_b(G \cdot b) \subset \xi_b$. Then $\dim V = \dim(G \cdot b)$. Since B is toric, $\dim B = 2 \dim G - 1$. Since

$$T_b B = \mathbb{R} Y_\alpha(b) \oplus V \oplus JV \oplus W \tag{III.1}$$

$\dim W = (2 \dim G - 1) - 1 - 2 \dim(G \cdot b) = 2 \dim G - 2(\dim G - \dim H) = 2 \dim H - 2$. Since (III.1) is a splitting as H-representations, the slice representation of H at b is $\mathbb{R} Y_\alpha(b) \oplus JV \oplus W$. As we observed earlier the action of H on $\mathbb{R} Y_\alpha(b) \oplus JV$ is trivial. Since the action of G on B is effective, the representation of H on W must be faithful. Contradiction.

Therefore $T_b(G \cdot b) \not\subset \xi_b$ and so $\dim V = \dim(G \cdot b) - 1$. In this case the dimension count gives us exactly that $\dim W = 2 \dim H$. $\qquad\square$

Lemma III.8. *Let $H \subset \mathbb{T}^2$ be a closed subgroup isomorphic to S^1. Then there is another closed subgroup $K \subset \mathbb{T}^2$ isomorphic to S^1 such that $\mathbb{T}^2 = K \times H$.*

Proof. Since H is isomorphic to S^1 it is of the form $\{\exp t\nu \mid t \in \mathbb{R}\}$ for some vector $\nu = (n_1, m_1) \in \mathbb{Z}^2 = \ker\{\exp : \mathbb{R}^2 \to \mathbb{T}^2\}$. We may assume that n_1 and

m_1 are relatively prime. Therefore there exist integers n_2, m_2 such that $n_1 m_2 - m_1 n_2 = 1$. Hence the vectors (n_1, m_1) and (n_2, m_2) form a basis of $\mathbb{Z}^2$. Take $K = \{\exp t(n_2, m_2) \mid t \in \mathbb{R}\}$. $\qquad\square$

Remark III.9. More generally if G is a torus and $H \subset G$ is a closed connected subgroup there is another closed connected subgroup $K \subset G$ so that $G = K \times H$. This is a bit harder to prove than the Lemma above.

III.2   The 3-dimensional case

In this section we prove Theorem I.38 in the case that dim $B = 3$:

Lemma III.10. *Let B be a compact connected contact toric $G = \mathbb{T}^2$ manifold (in particular dim $B = 3$). Suppose the action of G is not free. Then there exist two closed subgroups $K_1, K_2 \subset G$ isomorphic to S^1 so that B is homeomorphic to $([0,1] \times G)/ \sim$ where $(0, g) \sim (0, ag)$ for all $g \in G$, $a \in K_1$ and $(1, g) \sim (1, ag)$ for all $g \in G$ and $a \in K_2$. In other words B is obtained from the manifold with boundary $[0,1] \times G$ by collapsing circles in the two components of the boundary by the respective actions of two circle subgroups. (It may happen that $K_1 = K_2$.)*

Exercise III.11. Consider the standard sphere $S^3 = \{(z_1, z_2) \in C^2 \mid |z_1|^2 + |z_2|^2 = 1\}$. The torus $G = \{(\lambda_1, \lambda_2) \in C^2 \mid |\lambda_1|^2 = 1, \quad |\lambda_2|^2 = 1\}$ acts on S^3 by $(\lambda_1, \lambda_2) \cdot (z_1, z_2) = (\lambda_1 z_1, \lambda_2 z_2)$. Show that S^3 is of the form $([0,1] \times G)/ \sim$ for two circle subgroups K_1, K_2 of G where $\sim$ is the equivalence relation in Lemma III.10. What are the subgroups K_1, K_2?

Proof of Lemma III.10. By Lemma III.7 all isotropy groups for the action of G on B are connected and no isotropy group is all of G. Therefore, since dim $G = 2$ the possible isotropy groups are trivial or circles. And points with circle isotropy groups must exist since the action is not free.

If the isotropy group of a point $b \in B$ is trivial then (by Lemma III.2 and the dimension count) a G-invariant neighborhood U of G is equivariantly diffeomorphic to $G \times I$ where I is an open interval and G acts on $G \times I$ by $g \cdot (a, t) = (ga, t)$. Hence $U/G = I$ and there is a map $s : U/G \to B$ so that $\pi \circ s = id$ where $\pi : B \to B/G$ is the orbit map, i.e., π has a local section.

Now consider a point $b \in B$ with the isotropy group G_b isomorphic to S^1.[5] By Lemma III.2 a neighborhood of b in B is G-equivariantly diffeomorphic to a neighborhood of the zero section in $G \times_{G_b} W$. By Lemma III.8 there is a circle $L \subset G$ such that $G = G_b \times L$. Note that $G \cdot b \simeq L$ and that dim $W = 2$ and that by Lemma III.4 the representation of G_b on W is faithful. There is only one faithful real 2-dimensional representation of S^1 — it is the representation of S^1 on $\mathbb{R}^2$ as $SO(2)$ or, equivalently, the representation of S^1 on $\mathbb{C}$ as $U(1)$. Since $G = L \times G_b$, $G \times_{G_b} W = L \times W$, where $L \times G_b$ acts on $L \times W$ by $(\lambda, \mu) \cdot (g, w) =$

---

[5]By Lemma III.7 there are no more possibilities for G_b.

$(\lambda g, \mu \cdot w)$. We conclude that there is a G-invariant neighborhood U of $G \cdot b$ in M, an isomorphism $\varphi : G \to S^1 \times S^1$ and a diffeomorphism $\psi : U \to S^1 \times D^2$ such that $\psi(g \cdot x) = \varphi(g) \cdot \psi(x)$ for all $g \in G$, $x \in U$. Here $(\lambda, \mu) \in S^1 \times S^1 \subset \mathbb{C} \times \mathbb{C}$ acts on $S^1 \times D^2 \subset \mathbb{C} \times \mathbb{C}$ by

$$(\lambda, \mu) \cdot (z_1, z_2) = (\lambda z_1, \mu z_2).$$

Note that $S^1 \times D^2 = (\mathbb{T}^2 \times [0,1))/ \sim$ where $(\lambda_1, \lambda_2, 0) \sim (\lambda_1, \mu\lambda_2, 0)$ for all $(\lambda_1, \lambda_2) \in \mathbb{T}^2$ and $\mu \in S^1$. We conclude that

1. $U/G \simeq [0,1)$;

2. $U \simeq (G \times [0,1))/ \sim$ where $(g,0) \sim (ag,0)$ for all $g \in G$, $a \in G_b$;

3. there is a local section $s : U/G \to B$ of the orbit map $\pi : B \to B/G$;

4. the set of points in U with non-trivial isotropy groups is $G \cdot b = \pi^{-1}(0)$, where again $\pi : U \to [0,1)$ denotes the orbit map.

It follows that the orbit space B/G is locally homeomorphic to either an open interval or to a half-open interval $[0,1)$. Hence B/G is a topological 1-dimensional manifold with boundary. Since B/G is compact and connected we may identify it with $[0,1]$. Note that the set of points with non-trivial isotropy groups is $\pi^{-1}(\{0,1\})$, where by abuse of notation $\pi : B \to [0,1]$ denotes the orbit map. More specifically $\pi^{-1}(0) = G/K_1$, $\pi^{-1}(1) = G/K_2$ for some circle subgroups $K_1, K_2 \subset G$.

We now argue that $\pi : B \to [0,1]$ has a global section $s : [0,1] \to B$, so that $\pi \circ s(t) = t$ for all $t \in [0,1]$. We have seen that sections of π exist locally: for every $t \in [0,1]$ there is an interval $I \subset [0,1]$ open in $[0,1]$ and containing t and a map $s : I \to B$ so that $\pi \circ s = id_I$. We want to patch these local sections into a global section.

Since $[0,1]$ is compact, we can cover it by finitely many intervals I_j so that on each I_j there is a section $s_j : I_j \to B$. Let us now assume for simplicity that there are only two intervals: $I_0 = [0, 2/3)$ and $I_1 = (1/3, 1]$. The case of more than two intervals will be left as an exercise to the reader. Thus we have two sections $s_0 : [0, 2/3) \to B$ and $s_1 : (1/3, 1] \to B$. Since G acts freely on $\pi^{-1}((0,1))$ the map $\varphi : (1/3, 2/3) \times G \to \pi^{-1}((1/3, 2/3))$ given by $\varphi(t, g) = g \cdot s_0(t)$ is a G-equivariant diffeomorphism (where G acts on the product $(1/3, 2/3) \times G$ by multiplication on the second factor). We have $\varphi^{-1} \circ s_0(t) = (t, 1)$ and $\varphi^{-1} \circ s_1(t) = (t, g(t))$ for some curve $g : (1/3, 2/3) \to G$. Since $(1/3, 2/3)$ is simply connected we may lift g to a curve $\gamma(t)$ in the universal cover $\exp : \mathfrak{g} \to G$ of G. Choose a smooth function $\rho : (1/3, 2/3) \to [0,1]$ with $\rho(t) \equiv 0$ for t near $1/3$ and $\rho(t) \equiv 1$ for t near $2/3$. Now consider the curve $a : (1/3, 2/3) \to G$ given by $a(t) = \exp(\rho(t)\gamma(t))$. The map $s_2(t) = \varphi(t, a(t))$ is a local section of $\pi : B \to [0,1]$ which agrees with s_0 near $1/3$

and with s_1 near $2/3$. Thus we can define a global section $s : [0,1] \to B$ by

$$s(t) = \begin{cases} s_0(t) & t \in [0, 1/3], \\ s_2(t) & t \in [1/3, 2/3], \\ s_3(t) & t \in [1/3, 0]. \end{cases}$$

Now that we have a global section of $\pi : B \to [0,1]$ we can define a continuous map $\tilde{f} : [0,1] \times G \to B$, $\tilde{f}(t,g) = g \cdot s(t)$. The map is onto; it descends to a bijective continuous map $f : ([0,1] \times G)/ \sim \to B$ where $\sim$ is the equivalence relation in the statement of the Lemma. Since $([0,1] \times G)/ \sim$ is compact, the map f is a homeomorphism.   □

Exercise III.12. Show that if the groups K_1 and K_2 in the statement of the Lemma are the same, then B is $S^1 \times S^2$.

Exercise III.13. (This exercise is considerably harder than the one above.) Show that if the groups K_1 and K_2 are different, then B is the quotient of S^3 by a finite cyclic group.

Lemma III.14. *Let $G = \mathbb{T}^2$ and let $K_1, K_2 \subset G$ be two closed subgroups isomorphic to S^1. Let B be the topological space $([0,1] \times G)/ \sim$ where $(0,g) \sim (0,ag)$ for all $g \in G$, $a \in K_1$ and $(1,g) \sim (1,ag)$ for all $g \in G$ and $a \in K_2$. In other words B is obtained from the manifold with boundary $[0,1] \times G$ by collapsing circles in the two components of the boundary by the respective actions of two circle subgroups.*

Then either $H^1(B, \mathbb{Z}) = \mathbb{Z} = H^2(B, \mathbb{Z})$ or $H^1(B, \mathbb{Z}) = 0$ and $H^2(B, \mathbb{Z})$ is a finite group. In particular, B cannot be homeomorphic to the 3-torus $\mathbb{T}^3$.

Proof. Recall that $H^1(G, \mathbb{Z})$ is isomorphic to the weight lattice $\mathbb{Z}_G^*$ and that the isomorphism is given as follows: A weight $\nu \in \mathbb{Z}_G^*$ defines a character $\chi_\nu : G \to S^1$ by $\chi_\nu(\exp(X)) = e^{2\pi i \nu(X)}$; the class $\chi_\nu^*[d\theta]$ is the element in $H^1(G, \mathbb{Z})$ corresponding to ν. Here $d\theta$ is the obvious 1-form on S^1.

Consequently if $G = \mathbb{T}^2$ and $K_j \subset G$ is a circle subgroup, then $\pi_j : G \to G/K_j \simeq S^1$ is a character and hence the weight $\nu_j = (d\pi_j)_1$ defines an element of $H^1(G, \mathbb{Z})$. Thus if we identify $H^1(G/K_j, \mathbb{Z})$ with $\mathbb{Z}$ and $H^1(G, \mathbb{Z})$ with $\mathbb{Z}_G^*$, then the map $H^1(G/K_j, \mathbb{Z}) \to H^1(G, \mathbb{Z})$ becomes the map $\mathbb{Z} \ni n \mapsto n\nu_j \in \mathbb{Z}_G^*$.

The sets $U = ([0, 2/3) \times G)/ \sim$ and $V = ((1/3, 1] \times G)/ \sim$ are two open subsets of B. We have $B = U \cup V$, $U \cap V = (1/3, 2/3) \times G$ is homotopy equivalent to G, U is homotopy equivalent to G/K_1, V is homotopy equivalent to G/K_2 and the inclusion maps $U \cap V \hookrightarrow U$, $U \cap V \hookrightarrow V$ are homotopy equivalent to projections $\pi_1 : G \to G/K_1$, $\pi_2 : G \to G/K_2$ respectively. Hence under the above identifications of $H^1(U)$ and $H^1(V)$ with $\mathbb{Z}$, the inclusions $U \cap V \to U$, $U \cap V \to V$ induce the maps $\mathbb{Z} \ni n \mapsto n\nu_j \in \mathbb{Z}_G^*$, $j = 1, 2$, respectively.

We now apply the Mayer-Vietoris sequence to compute the integral cohomology of B. We have: $0 \to H^0(B) \to H^0(U) \oplus H^0(V) \to H^0(G) \xrightarrow{\delta} H^1(B) \to H^1(U) \oplus H^1(V) \to H^1(G) \xrightarrow{\delta} H^2(B) \to H^2(U) \oplus H^2(V) \to H^2(G) \xrightarrow{\delta} H^3(B) \to 0$.

Clearly the map $H^0(U) \oplus H^0(V) \to H^0(G)$ is onto. Given the identifications above the map $\varphi : H^1(U) \oplus H^1(V) \to H^1(G)$ becomes $\mathbb{Z} \oplus \mathbb{Z} \ni (n, m) \mapsto n\nu_1 + m\nu_2 \in \mathbb{Z}_G^*$. We therefore have $0 \to H^1(B) \to \mathbb{Z} \oplus \mathbb{Z} \xrightarrow{\varphi} \mathbb{Z}_G^* \xrightarrow{\delta} H^2(B) \to 0 \oplus 0 \to H^2(G) \xrightarrow{\delta} H^3(B) \to 0$. We conclude that

- $H^2(B) = \mathbb{Z}_G^* / (\mathbb{Z}\nu_1 + \mathbb{Z}\nu_2)$

- $H^1(B) = \{(n, m) \in \mathbb{Z}^2 \mid n\nu_1 + m\nu_2 = 0\}$.

Since ν_1, ν_2 are differentials of projections onto quotients by circle subgroups, either ν_1 and ν_2 are independent over $\mathbb{Z}$ or $\nu_1 = \pm\nu_2$. In the first case $H^1(B) = 0$ and $H^2(B)$ is a finite abelian group. In the second case $H^1(B) = \mathbb{Z}$ and $H^2(B) = \mathbb{Z}_G^* / \mathbb{Z}\nu_1 = \mathbb{Z}$. $\qquad\square$

This finishes the proof of Theorem I.38 in the case that $\dim B = 3$.

III.3   Uniqueness of symplectic toric manifolds

In this section we sketch a proof of

Theorem III.15. *Let $(B, \xi = \ker \alpha, \Psi : \xi_+^\circ \to \mathfrak{g}^*)$ be a compact connected contact toric G-manifold.*

Suppose the dimension k of the maximal linear subspace of the moment cone $C(\Psi) = \Psi(\xi_+^\circ) \cup \{0\}$ satisfies $0 < k < \dim G$. Then B is homotopy equivalent to the product of a k-torus with a sphere. In particular B is not the co-sphere bundle of the n-torus, $n = \frac{1}{2}(\dim B + 1) = \dim G$.

The main idea of the proof is simple. We will first argue that there is an action of G on the symplectic manifold

$$M = T^*\mathbb{T}^k \times \mathbb{C}^l \setminus 0 = \{(q, p, z) \in \mathbb{T}^k \times (\mathbb{R}^k)^* \times \mathbb{C}^l = T^*\mathbb{T}^k \times \mathbb{C}^l \mid (p, z) \neq (0, 0)\},$$

$l = \frac{1}{2}(\dim B + 1 - k) > 0$, with moment map $\tilde{\Phi} : M \to \mathfrak{g}^*$ such that

$$\tilde{\Phi}(M) = \Psi(\xi_+^\circ). \tag{III.2}$$

We then argue that (III.2) implies that M is G-equivariantly symplectomorphic to ξ_+°. Note that M is homotopy equivalent to $\mathbb{T}^k \times S^{k+2l-1}$.

We start with the definition of a symplectic slice representation (c.f. proof of Lemma III.7), which is essential for understanding the local structure of symplectic toric manifolds.

Definition III.16. Let (M, ω) be a symplectic manifold with a Hamiltonian action of a torus G. Then an orbit $G \cdot m$ is an isotropic submanifold of (M, ω).[6] The **symplectic slice representation** at m is the representation of the isotropy group

---

[6]For a proof of this easy fact see, for example, [GS].

G_m of m on the symplectic vector space $V := T_m(G \cdot m)^\omega / T_m(G \cdot m)$. Here, as usual, $T_m(G \cdot m)^\omega$ denotes the symplectic perpendicular to $T_m(G \cdot m)$ in the symplectic vector space $(T_m M, \omega_m)$.

The equivariant isotropic embedding theorem (see for example [GS], Theorem 39.1) asserts that a neighborhood of an orbit $G \cdot m$ is determined (up to equivariant symplectomorphisms) by the symplectic slice representation at m. In fact, the topological normal bundle of the orbit $G \cdot m$ in M is $G \times_H (\mathfrak{g}/\mathfrak{h} \times V)$ where H is the isotropy group of m and $\mathfrak{h}$ is its Lie algebra (*op. cit.*).

Remark III.17. In the case of symplectic toric manifolds the dimension of the symplectic slice V at m is twice the dimension of the isotropy group G_m. Hence by Lemma III.6 the group G_m is connected and the representation of G_m on V is determined by a set of weights $\{\nu_i\}$ which forms a basis of the weight lattice of G_m. Note that the image of V under the moment map $\Phi_V : V \to \mathfrak{g}_m^*$ defined by the representation is the cone

$$\Phi_V(V) = \{\sum a_i \nu_i \mid a_i \geq 0\}.$$

In particular the edges of the cone are spanned by the weights. Alternatively, the isotropy group G_m is isomorphic to $\mathbb{T}^l$ for some l and the slice representation $\rho : G_m \to \mathrm{Sp}(V)$ is isomorphic to the standard representation of $\mathbb{T}^l$ on $\mathbb{C}^l$.[7]

With a little more work one can prove the following two propositions (their proofs can be found, for example, in [D]).

Proposition III.18. *Let* $(M, \omega, \Phi : M \to \mathfrak{g}^*)$ *be a symplectic toric manifold,* $m \in M$ *a point and* U *a neighborhood of* $G \cdot m$ *in* M. *Then for a sufficiently small ball* $\mathcal{O}$ *about* $\eta = \Phi(m)$ *in* $\mathfrak{g}^*$ *the set* $\Phi(U) \cap \mathcal{O}$ *determines the symplectic slice representation at* m *and hence a small* G-*invariant neighborhood of* $G \cdot m$ *in* M.

Proposition III.19. *Let* $(M, \omega, \Phi : M \to \mathfrak{g}^*)$ *be a symplectic toric manifold,* $m \in M$ *a point,* $\eta = \Phi(m)$ *and* G_m *the isotropy group of* m. *Identify* G_m *with the standard torus* $\mathbb{T}^d$, $d = \dim G_m$, *and extend it to an identification of* G *with* $\mathbb{T}^d \times \mathbb{T}^c$, [8] $c = \dim G - \dim G_m$.

A G-*invariant neighborhood* U *of* $G \cdot m$ *in* M *is equivariantly symplectomorphic to a neighborhood of* $\mathbb{T}^c \times \{(0,0)\}$ *in* $T^*\mathbb{T}^c \times \mathbb{C}^d \simeq \mathbb{T}^c \times (\mathbb{R}^c)^* \times \mathbb{C}^d$. *Hence*

$$\{t(\mu - \eta) + \eta \mid t \geq 0, \, \mu \in \Phi(U)\} = \tilde{\Phi}(T^*\mathbb{T}^c \times \mathbb{C}^d),$$

where $\tilde{\Phi} : T^*\mathbb{T}^c \times \mathbb{C}^d \to (\mathbb{R}^c)^* \times (\mathbb{R}^d)^* \simeq \mathfrak{g}^*$ *is the moment map for the "obvious" action of* $G = \mathbb{T}^c \times \mathbb{T}^d$ *on* $T^*\mathbb{T}^c \times \mathbb{C}^d$.[9]

---

[7] The standard representation of $\mathbb{T}^l = \{(\lambda_1, \ldots, \lambda_l) \in \mathbb{C}^l \mid |\lambda_j| = 1\}$ on $\mathbb{C}^l$ is given by $(\lambda_1, \ldots, \lambda_l) \cdot (z_1, \ldots, z_l) = (\lambda_1 z_1, \ldots, \lambda_l z_l)$.

[8] Here we use Remark III.9.

[9] The "obvious" action of $\mathbb{T}^c$ on $T^*\mathbb{T}^c$ is the lift of left multiplication.

Corollary III.20. *Let* $(B, \xi = \ker \alpha, \Psi : \xi_+^\circ)$ *be a compact connected contact toric G-manifold. Suppose the dimension k of the maximal linear subspace of the moment cone $C(\Psi) = \Psi(\xi_+^\circ) \cup \{0\}$ satisfies $0 < k < \dim G$.*
Then there is an identification of G with $\mathbb{T}^k \times \mathbb{T}^l$, $l = \dim G - k$, so that

$$\Psi(\xi_+^\circ) \cup \{0\} = \tilde{\Phi}(T^*\mathbb{T}^k \times \mathbb{C}^l)$$

where $\tilde{\Phi} : T^*\mathbb{T}^k \times \mathbb{C}^l \to (\mathbb{R}^k)^* \times (\mathbb{R}^l)^* \simeq \mathfrak{g}^*$ is the moment map for the obvious action of $\mathbb{T}^k \times \mathbb{T}^l$ on $T^*\mathbb{T}^k \times \mathbb{C}^l$.

Proof. If C is a cone in $\mathfrak{g}^*$ whose maximal linear subspace is P and if U is a neighborhood in $\mathfrak{g}^*$ of a point $\eta \in P$ then C equals the cone on $U \cap C$ with the vertex at η:

$$C = \{t(\mu - \eta) + \eta \mid t \geq 0, \mu \in U \cap C\}.$$

Since the B is contact toric $\Psi(\xi_+^\circ)$ does not contain the origin (c.f. Lemma III.7). Since B is compact and $\Psi : \xi_+^\circ \to \mathfrak{g}^*$ is homogeneous, $\Psi : \xi_+^\circ \to \mathfrak{g}^* \smallsetminus \{0\}$ is proper. Hence for any $\eta \in \Psi(\xi_+^\circ)$ and any neighborhood U of $\Psi^{-1}(\eta)$, $\Psi(U)$ is a neighborhood of η in $\Psi(\xi_+^\circ)$.

Moreover, since the fibers of Ψ are G-orbits (by Theorem III.1), for any neighborhood U of an orbit $G \cdot m$ the set $\Psi(U)$ is a neighborhood of $\eta = \Psi(m)$ in $\Psi(\xi_+^\circ)$. In particular it contains a set of the form $\mathcal{O} \cap \Psi(\xi_+^\circ)$ where $\mathcal{O} \subset \mathfrak{g}^*$ is a small ball about η.

Let $P \subset C(\Psi)$ be the maximal linear subspace and let $0 \neq \eta \in P$. Then, as noted at the beginning of the proof, for any ball $\mathcal{O} \subset \mathfrak{g}^*$ about η we have

$$C(\Psi) = \{t(\mu - \eta) + \eta \mid t \geq 0, \mu \in \mathcal{O} \cap \Psi(\xi_+^\circ)\}$$
$$= \{t(\mu - \eta) + \eta \mid t \geq 0, \mu \in \Psi(U)\}.$$

On the other hand, it follows from Proposition III.19 that

$$\{t(\mu - \eta) + \eta \mid t \geq 0, \mu \in \Psi(U)\} = \tilde{\Phi}(T^*\mathbb{T}^c \times \mathbb{C}^d) \simeq \mathbb{R}^c \times (\mathbb{R}_{\geq 0})^d$$

for some integers c and d. It follows that $c = k$, $d = \dim G - k = l$ and that

$$C(\Psi) = \tilde{\Phi}(T^*\mathbb{T}^k \times \mathbb{C}^l).$$

$\square$

Note that since $\tilde{\Phi}^{-1}(0) = \mathbb{T}^k \times \{(0,0)\}$, we get

$$\Psi(\xi_+^\circ) = \tilde{\Phi}(M),$$

where $M = (T^*\mathbb{T}^k \times \mathbb{C}^l) \smallsetminus (\mathbb{T}^k \times \{(0,0)\})$.

Definition III.21. A symplectic toric G-manifold $(M, \omega, \Phi : M \to \mathfrak{g}^*)$ is **good** (for the purposes of these lectures) if

1. the fibers of $\Phi : M \to \mathfrak{g}^*$ are connected,

2. the set $\Phi(M)$ is contractible and

3. there is an open set $U \subset \mathfrak{g}^*$ such that $\Phi(M) \subset U$ and the map $\Phi : M \to U$ is proper.

The definition is designed in such a way that it includes compact symplectic toric manifolds, symplectizations of compact contact toric manifolds of dimension bigger than 3 and the manifolds of the form $M = (T^*\mathbb{T}^k \times \mathbb{C}^l) \smallsetminus (\mathbb{T}^k \times \{(0,0)\})$, $0 < k, l$. As a consequence of the definition and of Proposition III.18 we have

Lemma III.22. *Let* $(M, \omega, \Phi : M \to \mathfrak{g}^*)$ *be a good symplectic toric manifold. For any point* $\eta \in \Phi(M)$ *and any sufficiently small ball* $\mathcal{O}$ *centered at* η, *the set* $\mathcal{O} \cap \Phi(M)$ *determines the symplectic toric manifold* $(\Phi^{-1}(\mathcal{O}), \omega|_{\Phi^{-1}(\mathcal{O})}, \Phi|_{\Phi^{-1}(\mathcal{O})})$ *(up to a* G-*equivariant symplectomorphism).*

Definition III.23. Two symplectic toric G-manifolds $(M, \omega, \Phi : M \to \mathfrak{g}^*)$ and $(M', \omega', \Phi' : M' \to \mathfrak{g}^*)$ are **isomorphic** if there is a G-equivariant diffeomorphism $\sigma : M \to M'$ such that $\sigma^*\omega' = \omega$ and $\sigma^*\Phi' = \Phi$.

We denote by $\mathrm{Iso}(M, \omega, \Phi) = \mathrm{Iso}(M)$ the group of isomorphisms of a symplectic toric manifold (M, ω, Φ).

Note that the last condition on σ is almost redundant — if σ is symplectic and G-equivariant then $\sigma^*\Phi' = \Phi + c$ for some constant vector $c \in \mathfrak{g}^*$. We impose the last condition for technical convenience.

Definition III.24. Two symplectic toric G-manifolds $(M, \omega, \Phi : M \to \mathfrak{g}^*)$ and $(M', \omega', \Phi' : M' \to \mathfrak{g}^*)$ are **locally isomorphic** over a set $\Delta \subset \mathfrak{g}^*$ if

1. $\Phi(M) = \Delta = \Phi'(M')$ and

2. for any $\eta \in \Delta$ and any sufficiently small ball $\mathcal{O} \subset \mathfrak{g}^*$ centered at η the symplectic toric manifolds $(\Phi^{-1}(\mathcal{O}), \omega|_{\Phi^{-1}(\mathcal{O})}, \Phi|_{\Phi^{-1}(\mathcal{O})})$ and $((\Phi')^{-1}(\mathcal{O}), \omega'|_{(\Phi')^{-1}(\mathcal{O})}, \Phi'|_{(\Phi')^{-1}(\mathcal{O})})$ are isomorphic.

Given the above definitions we see that Lemma III.22 implies

Lemma III.25. *Suppose* (M, ω, Φ) *and* (M', ω', Φ') *are two good symplectic toric* G-*manifolds with* $\Phi(M) = \Phi'(M')$. *Then* (M, ω, Φ) *and* (M', ω', Φ') *are locally isomorphic over* $\Delta = \Phi(M)$.

Therefore the proof of Theorem III.15 reduces to

Proposition III.26. *Any good symplectic toric* G-*manifold locally isomorphic to a given symplectic toric* G-*manifold* (M, ω, Φ) *is actually isomorphic to* (M, ω, Φ).

The proposition could be stated for a larger class of symplectic toric manifolds. We leave it to the reader to find the most general form of the statement (and prove it). The rest of the section is occupied with a proof of the proposition.

Suppose a symplectic toric G-manifold (M', ω', Φ') is locally isomorphic to (M, ω, Φ). Then for any $\eta \in \Phi(M)$ and for any sufficiently small ball $\mathcal{O} \subset \mathfrak{g}^*$ about η the symplectic toric manifold $\Phi^{-1}(\mathcal{O})$ is isomorphic to $(\Phi')^{-1}(\mathcal{O})$. Choose a locally finite cover $\{\mathcal{O}_i\}_{i \in I}$ of $\Phi(M)$ by such balls. Then for each index i we have an isomorphism

$$f_i : \Phi^{-1}(\mathcal{O}_i) \rightarrow (\Phi')^{-1}(\mathcal{O}_i).$$

Let $\mathcal{O}_{ij} = \mathcal{O}_i \cap \mathcal{O}_j$. Define

$$g_{ij} : \Phi^{-1}(\mathcal{O}_{ij}) \rightarrow \Phi^{-1}(\mathcal{O}_{ij}), \quad g_{ij} = f_i^{-1} \circ f_j$$

(to keep the notation manageable we wrote f_i for the restriction $f_i|_{\mathcal{O}_{ij}}$ etc.; we will continue to omit restrictions in the rest of the section). It is easy to see that

$$g_{ii} = id, \ g_{ij} \circ g_{ji} = id, \text{ and } \ g_{ij} \circ g_{jk} \circ g_{ki} = id \qquad \text{(III.3)}$$

wherever these equations make sense.

The data $\{\mathcal{O}_i\}$, $\{g_{ij}\}$ and (M, ω, Φ) allow us to reconstruct (M', ω', Φ'). Indeed, let $\tilde{M} = \bigsqcup \Phi^{-1}(\mathcal{O}_i)$. Define a relation $\sim$ on $\tilde{M}$ by $\Phi^{-1}(\mathcal{O}_i) \ni x_i \sim x_j \in \Phi^{-1}(\mathcal{O}_j)$ iff $x_j = g_{ij}(x_i)$. Equations (III.3) imply that $\sim$ is an equivalence relation. It follows that $\tilde{M}/\sim$ is a symplectic toric manifold. Moreover the map $\tilde{F} : \tilde{M} \rightarrow M'$ defined by $\tilde{F}|_{\Phi^{-1}(\mathcal{O}_i)} = f_i$ descends to a well defined map $F : \tilde{M}/\sim \rightarrow M'$. The map F is an isomorphism.

Suppose $\tilde{f}_i : \Phi^{-1}(\mathcal{O}_i) \rightarrow (\Phi')^{-1}(\mathcal{O}_i)$ is another collection of isomorphisms. Let $\tilde{g}_{ij} = (\tilde{f}_i)^{-1} \circ \tilde{f}_j$. Clearly

$$g_{ij} = h_i \circ \tilde{g}_{ij} \circ h_j^{-1} \qquad \text{(III.4)}$$

where $h_i = f_i^{-1} \circ \tilde{f}_i \in \text{Iso}(\Phi^{-1}(\mathcal{O}_i))$.

We also get a different set of data if we choose a different cover. However, all these sets of data define one object — a class in the first Čech cohomology with coefficients in a certain sheaf. Let us now review the notions of sheaves and Čech cohomology. There are many good references for this material. I will be following [WW].

Definition III.27. A **sheaf** of groups $\mathcal{S}$ on a topological space X is an assignment

$$\mathcal{S} : \{\text{opens sets in } X\} \rightarrow \text{groups}, \ U \mapsto \mathcal{S}(U)$$

satisfying two conditions:

1. For a pair of open sets $U \subset W$ in X there is a restriction map $\rho_U^W : \mathcal{S}(W) \rightarrow \mathcal{S}(U)$ such that for any three sets $U \subset W \subset V$ of X

$$\rho_U^W \circ \rho_W^V = \rho_U^V.$$

Elements of $\mathcal{S}(U)$ are called **sections**. Given a section $\varphi \in \mathcal{S}(W)$ we write $\varphi|_U$ for $\rho_U^W(\varphi)$.

2. Given an open cover $\{U_i\}$ of an open set U (so that $U = \bigcup U_i$) and a collection of sections $\varphi_i \in \mathcal{S}(U_i)$ such that

$$\varphi_i|_{U_i \cap U_j} = \varphi_j|_{U_i \cap U_j}$$

for all indices i, j there is a unique section $\varphi \in \mathcal{S}(U)$ such that

$$\varphi|_{U_i} = \varphi_i$$

for all i.

The sheaf $\mathcal{S}$ is **abelian** if $\mathcal{S}(U)$ is an abelian group for all open sets U.

Three examples of sheaves will be important to us. Check that they are indeed sheaves.

Example III.28. Let (M, ω, Φ) be a good symplectic toric G-manifold. The assignment Iso : $U \mapsto \text{Iso}(U)$ ($U \subset \Phi(M)$ open) is a sheaf on $\Phi(M)$. The group operation is composition.

Example III.29 *(Locally constant sheaf)*. Let H be a group and X a topological space. The assignment that associates the group H to every open connected subset of X is a sheaf, called a locally constant sheaf. It is denoted by $\underline{H}$. Thus $\underline{H}(U) = H$ for every connected open set $U \subset X$. The group operation is the multiplication in H.

Example III.30. Let (M, ω, Φ) be a good symplectic toric G-manifold. Define a sheaf $\mathcal{C}$ on $\Phi(M)$ by

$$\mathcal{C}(U) = C^\infty(\Phi^{-1}(U))^G, \quad G\text{-invariant smooth functions on } \Phi^{-1}(U).$$

The group operation is addition of functions.

Definition III.31. Let $\mathcal{S}_1, \mathcal{S}_2$ be sheaves on a topological space X. A **map of sheaves** $\tau : \mathcal{S}_1 \to \mathcal{S}_2$ is a family of group homomorphisms

$$\tau_U : \mathcal{S}_1(U) \to \mathcal{S}_2(U), \quad U \subset X \text{ open}$$

compatible with the restrictions:

$$(\rho_2)_U^W \circ f_W = f_U \circ (\rho_1)_U^W \quad \text{for all pairs } U \subset W \text{ of open sets in } X.$$

Example III.32. Consider the sheaves Iso and $\mathcal{C}$ defined in Examples III.28 and III.30 above. A section $f \in \mathcal{C}(U)$ is a G-invariant function on $\Phi^{-1}(U)$. Its time t flow φ_t^f preserves the fibers of Φ and is a G-equivariant symplectomorphism of $\Phi^{-1}(U)$. Hence φ_1^f is a section of Iso(U). This gives us for each open set $U \subset \Phi(M)$ a map $\tau_U : \mathcal{C}(U) \to \text{Iso}(U)$, $\tau_U(f) = \varphi_1^f$. Moreover, any two functions $f_1, f_2 \in \mathcal{C}(U)$ Poisson commute [prove this]. Hence $\varphi_t^{f_1} \circ \varphi_t^{f_2} = \varphi_t^{f_1+f_2}$ and therefore τ_U is a group homomorphism. Thus we get a map of sheaves $\tau : \mathcal{C} \to \text{Iso}$.

Given a map of sheaves $\tau : \mathcal{S}_1 \to \mathcal{S}_2$ one can define the sheaves the kernel and image sheaves $\ker \tau$ and $\mathrm{im}\tau$: for an open set U $(\ker \tau)(U) = \ker \tau_U$, $(\mathrm{im}\,\tau)(U) = \mathrm{im}\,\tau U$. Hence it makes sense to say that a map of sheaves is onto and more generally talk about exact sequences of sheaves.[10]

Proposition III.33. *Let* (M, ω, Φ) *be a good symplectic toric manifold. The map of sheaves* $\tau : \mathcal{C} \to \mathrm{Iso}$ *defined above is onto. Hence* Iso *is an abelian sheaf.*

The kernel of τ *is the locally constant sheaf* $\underline{\mathbb{R} \times \mathbb{Z}_G}$. *We thus have a short exact sequence of abelian sheaves:*

$$0 \to \underline{\mathbb{R} \times \mathbb{Z}_G} \to \mathcal{C} \to \mathrm{Iso} \to 0.$$

The proposition is due to Boucetta and Molino [BoM]. See also [LT].

III.3.1   Čech cohomology

In this subsection we "review" the notion of Čech cohomology with coefficients in an abelian sheaf. There are many good references, such as [WW], for the nontrivial facts that we list below without proofs.

Let X be a topological space $\{U_i\}$ an open locally finite cover of X and $\mathcal{S}$ an abelian sheaf on X. A 0 Čech cochain is a function that assigns to each index i an element f_i of $\mathcal{S}(U_i)$, i.e., the group of 0-cochains $C^0(\{U_i\}, \mathcal{S})$ is the product $\prod \mathcal{S}(U_i)$. A 1 Čech cochain assigns to an ordered pair of indices ij an element g_{ij} of $\mathcal{S}(U_{ij})$ where $U_{ij} = U_i \cap U_j$. Moreover we require that $g_{ij} = -g_{ji}$ (we now think of the groups $\mathcal{S}(U)$ additively). More generally a p-cochain assigns to an ordered $p + 1$ tuple of indices $i_0 \ldots i_p$ an element $s_{i_0 \ldots i_p} \in \mathcal{S}(U_{i_0 \ldots i_p})$ where $U_{i_0 \ldots i_p} = U_{i_0} \cap \ldots \cap U_{i_p}$ and $s_{i_0 \ldots i_p}$ is skew-symmetric in the indices. The coboundary operator $\delta : C^p(\{U_i\}, \mathcal{S}) \to C^{p+1}(\{U_i\}, \mathcal{S})$ is defined by

$$(\delta s)_{i_0 \ldots i_{p+1}} = \sum (-1)^j s_{i_0 \ldots \hat{i}_j \ldots i_{p+1}}$$

where $\hat{i}_j$ means that the index is omitted, and where we omitted writing the restrictions of the terms on the right hand side to $U_{i_0 \ldots i_{p+1}}$. One proves that $\delta^2 = 0$. The cohomology of the complex $(C^p(\{U_i\}, \mathcal{S}), \delta)$ denoted by $\check{H}^*(\{U_i\}, \mathcal{S})$ is called the Čech cohomology of the cover $\{U_i\}$ with coefficients in the sheaf $\mathcal{S}$.

Given a refinement $\{V_j\}$ of the cover $\{U_i\}$ the restrictions give rise to a chain map $C^p(\{U_i\}, \mathcal{S}) \to C^p(\{V_j\}, \mathcal{S})$, which in turn gives rise to a map in cohomology $\check{H}^*(\{U_i\}, \mathcal{S}) \to \check{H}^*(\{V_j\}, \mathcal{S})$. Taking the direct limit over all locally finite covers we get a well-defined cohomology group

$$\check{H}^*(X, \mathcal{S}) = \varinjlim \check{H}^*(\{U_i\}, \mathcal{S}),$$

the Čech cohomology of X with coefficients in the sheaf $\mathcal{S}$.

---

[10]Warning: the map $\tau : \mathcal{S}_1 \to \mathcal{S}_2$ being onto *does not* mean that τ_U is onto for every open set U. See [WW] or any other good book on sheaves for more details.

Now let $(M, \omega, \Phi : M \to \mathfrak{g}^*)$ be a good symplectic toric G-manifold and $\{\mathcal{O}_i\}$ a locally finite cover of $\Phi(M)$ by sufficiently small balls. If $\{g_{ij}\} \in C^1(\{\mathcal{O}_i\}, \mathrm{Iso})$ is a 1-cochain, then $\delta(\{g_{**}\}) = 0$ means that for all triples of indices ijk we have

$$0 = \delta(\{g_{**}\})_{ijk} = -g_{jk} + g_{ik} + g_{ij},$$

which is (III.3) in additive notation (where on the right hand side we omitted the restrictions to U_{ijk}). Similarly if $\{g_{ij}\}, \{\tilde{g}_{ij}\} \in C^1(\{\mathcal{O}_i\}, \mathrm{Iso})$ are two 1-cochains that differ by $\delta(\{h_*\})$ for some 0-cochain $\{h_i\}$ then

$$\tilde{g}_{ij} - g_{ij} = -h_i + h_j$$

hence

$$g_{ij} = h_i + \tilde{g}_{ij} - h_j,$$

which is (III.4) in additive notation. Thus the discussion above shows that to every element of $\check{H}^1(\{\mathcal{O}_i\}, \mathrm{Iso})$ there corresponds a good symplectic toric manifold $(M', \omega', \Phi' : M \to \mathfrak{g}^*)$ locally isomorphic to $(M, \omega, \Phi : M \to \mathfrak{g}^*)$. More generally one can check that there is a one-to-one correspondence between cohomology classes in $\check{H}^1(\Phi(M), \mathrm{Iso})$ and isomorphism classes of good symplectic toric manifolds locally isomorphic to $(M, \omega, \Phi : M \to \mathfrak{g}^*)$. Thus to complete the proof of Proposition III.26 (and thereby Theorem III.15) it remains to show that the group $\check{H}^1(\Phi(M), \mathrm{Iso})$ is trivial for any good symplectic toric manifold M. For this we use Proposition III.33, two properties of Čech cohomology and a property of the sheaf $\mathcal{C}$ defined in Example III.30. The first property of Čech cohomology that we need is

Theorem III.34. *A short exact sequence of abelian sheaves $0 \to \mathcal{S}_1 \to \mathcal{S}_2 \to \mathcal{S}_3 \to 0$ on a space X induces a long exact sequence in Čech cohomology*

$$\cdots \to \check{H}^p(X, \mathcal{S}_1) \to \check{H}^p(X, \mathcal{S}_2) \to \check{H}^p(X, \mathcal{S}_3) \xrightarrow{\delta} \check{H}^{p+1}(X, \mathcal{S}_1) \to \cdots$$

The second property that we will use is

Theorem III.35. *Let X be a simply connected topological space and H an abelian group. The Čech cohomology $\check{H}^*(X, \underline{H})$ of X with coefficients in the locally constant sheaf $\underline{H}$ is isomorphic to the singular cohomology $H^*(X, H)$ of X with coefficients in the abelian group H.*

We will use the following property of the sheaf $\mathcal{C}$ (cf. [LT], Proposition 7.3)

Lemma III.36. *The sheaf $\mathcal{C}$ defined in Example III.30 is acyclic, that is,* $\check{H}^q(\Phi(M), \mathcal{C}) = 0$ *for all $q > 0$.*

Now putting Theorem III.34, Lemma III.36 and Proposition III.33 together we see that if (M, ω, Φ) is a good symplectic toric G-manifold and Iso the sheaf defined in Example III.28 then the cohomology group $\check{H}^1(\Phi(M), \mathrm{Iso})$ is isomorphic to $\check{H}^2(\Phi(M), \mathbb{R} \times \mathbb{Z}_G)$. The latter group is isomorphic to the singular cohomology group $H^2(\Phi(M), \mathbb{Z}_G \times \mathbb{R})$ by Theorem III.35. But $\Phi(M)$ is contractible, so $H^2(\Phi(M), \mathbb{Z}_G \times \mathbb{R}) = 0$. Therefore $\check{H}^1(\Phi(M), \mathrm{Iso}) = 0$, which proves Proposition III.26 and thereby Theorem III.15.

III.4 Proof of Theorem I.38, part three

The goal of this section is to prove

Theorem III.37. *Let* $(B, \xi = \ker \alpha, \Psi : \xi^\circ_+ \to \mathfrak{g}^*)$ *be a compact connected contact toric G-manifold with* $\dim B \geq 3$. *Suppose there is a vector* X *in the Lie algebra* $\mathfrak{g}$ *of G such that the function* $\langle \Psi, X \rangle$ *is strictly positive on B. Then* $\dim H^1(B, \mathbb{R}) \leq 1$. *In particular B is not the co-sphere bundle* $S^*G = \mathbb{T}^n \times S^{n-1}$, $n = \dim G = \frac{1}{2}(\dim B + 1) \geq 2$.

The proof of Theorem III.37 above will complete our proof of Theorem I.38. Since Theorem I.38 implies the main result of the notes, Theorem I.4, this, in turn, will finish the proof of the main result. As was sketched out at the beginning of Chapter 4 our proof of Theorem III.37 has several steps. The first one is a theorem implicit in a paper of Boyer and Galicki [BG]:

Theorem III.38. *Let* $(B, \xi = \ker \alpha, \Psi : \xi^\circ_+ \to \mathfrak{g}^*)$ *be a compact connected contact toric G-manifold with* $\dim B \geq 3$. *Suppose there is a vector* X *in the Lie algebra* $\mathfrak{g}$ *of G such that the function* $\langle \Psi, X \rangle$ *is strictly positive on B. Then there exists on B a locally free circle action so that the quotient* $M = B/S^1$ *is a (compact) symplectic toric orbifold.*

The second step is the argument that if M is a compact connected symplectic toric orbifold then $H^q(M, \mathbb{R}) = 0$ for all *odd* degrees q. This step uses Morse theory on orbifolds.

Let us now see why these two steps give us a proof of Theorem III.37. Consider the circle action produced by Theorem III.38 and the corresponding S^1 orbit map $\pi : B \to M$. If the circle action is actually free, then π is a circle fibration and we have the Gysin sequence

$$0 = H^{-1}(M, \mathbb{R}) \to H^1(M, \mathbb{R}) \to H^1(B, \mathbb{R}) \to H^0(M, \mathbb{R}) \to H^2(M, \mathbb{R}) \to \cdots .$$
(III.5)

If the action of S^1 is locally free, the long exact sequence (III.5) still exists. The reason is that the Gysin sequence arises from a collapse of the Leray-Serre spectral sequence for a sphere bundle. For locally free S^1 actions the orbit map $\pi : B \to M$ is not a fibration, but the corresponding spectral sequences still collapses if we use real coefficients.[11] Now, since $H^1(M, \mathbb{R}) = 0$ by the second step, it follows from (III.5) that $\dim H^1(B, \mathbb{R}) \leq \dim H^0(M, \mathbb{R}) = 1$, which proves Theorem III.37.

Here is an outline of the rest of the section. We start with a review of the notion of orbifold. We recall how orbifolds arise as quotients of manifolds by locally free actions of compact Lie groups. We then review the symplectic reduction theorem, in particular the fact that generically symplectic reduced spaces are orbifolds and use it to prove Theorem III.38. Next we define Morse functions on orbifolds, and discuss the two fundamental results of Morse theory on orbifolds. Finally we

---

[11] The reader not familiar with spectral sequences may wish to take this claim on faith.

argue that a compact symplectic toric orbifold M has a Morse function with all indices even and use this to conclude that the real cohomology of M vanishes in odd degrees.

We now define orbifolds and related differential geometric notions. For more details, see Satake [Sa1, Sa2] and, for a more modern point of view, Ruan [R]. The notion of orbifold was introduced by Satake in 1956 under the name of V-manifold. Orbifolds are designed to generalize manifolds in the following sense: an n-dimensional manifold is locally modeled on an open subset of $\mathbb{R}^n$. An n-dimensional orbifold is locally modeled on a quotient $\tilde{U}/\Gamma$ where $\tilde{U}$ is an open subset of $\mathbb{R}^n$ and Γ is a finite group acting smoothly on $\tilde{U}$. To give a precise definition we need a few preliminary notions — we need to define charts, atlases and compatibility of atlases.

Let U be a connected topological space, $\tilde{U}$ a connected n-dimensional manifold and Γ a finite group acting smoothly on $\tilde{U}$. An n-**dimensional uniformizing chart**[12] on U is a triple $(\tilde{U}, \Gamma, \varphi)$ where $\varphi : \tilde{U} \to U$ is a continuous map inducing a homeomorphism between $\tilde{U}/\Gamma$ and U (thus, in particular, φ is constant on the orbits of Γ). We will only consider charts where the set of points in $\tilde{U}$ fixed by Γ is either all of $\tilde{U}$ or is of codimension 2 or greater. Note that we **do not** require that Γ acts effectively.

Two uniformizing charts $(\tilde{U}_1, \Gamma_1, \varphi_1)$ and $(\tilde{U}_2, \Gamma_2, \varphi_2)$ of U are **isomorphic** if there is a diffeomorphism $\psi : \tilde{U}_1 \to \tilde{U}_2$ and an isomorphism $\lambda : \Gamma_1 \to \Gamma_2$ such that ψ is λ-equivariant (i.e., $\psi(g \cdot x) = \lambda(g) \cdot \psi(x)$ for all $g \in \Gamma_1$, $x \in \tilde{U}_1$) and $\varphi_2 \circ \psi = \varphi_1$. For example, fix $a \in \Gamma$. Define $\psi : \tilde{U} \to \tilde{U}$ by $\psi(x) = a \cdot x$. Let $\lambda(g) = aga^{-1}$. Then $(\psi, \lambda) : (\tilde{U}, \Gamma, \varphi) \to (\tilde{U}, \Gamma, \varphi)$ is an isomorphism.

Let $\iota : U' \hookrightarrow U$ be a connected open subset of U. We say that a uniformizing chart $(\tilde{U}', \Gamma', \varphi')$ is **induced from** a uniformizing chart $(\tilde{U}, \Gamma, \varphi)$ on U if there is a monomorphism $\lambda : \Gamma' \to \Gamma$ and a λ-equivariant embedding $\psi : \tilde{U}' \to \tilde{U}$ such that $\iota \circ \varphi' = \varphi \circ \psi$. In this case $(\psi, \lambda) : (\tilde{U}', \Gamma', \varphi') \to (\tilde{U}, \Gamma, \varphi)$ is called an **injection**.

An **orbifold atlas** on a Hausdorff topological space M is an open cover $\mathcal{U}$ of M satisfying the following conditions:

1. Each element U of $\mathcal{U}$ is uniformized, say by $(\tilde{U}, \Gamma, \varphi)$.

2. If $U, U' \in \mathcal{U}$ and $U' \subset U$ then there is an injection $(\tilde{U}', \Gamma', \varphi') \to (\tilde{U}, \Gamma, \varphi)$.

3. For any point $p \in U_1 \cap U_2$, $U_1, U_2 \in \mathcal{U}$, there is a connected open set $U_3 \in \mathcal{U}$ with $p \in U_3 \subset U_1 \cap U_2$ (and hence there are injections $(\tilde{U}_3, \Gamma_3, \varphi_3) \to (\tilde{U}_1, \Gamma_1, \varphi_1)$ and $(\tilde{U}_3, \Gamma_3, \varphi_3) \to (\tilde{U}_2, \Gamma_2, \varphi_2)$).

Suppose that $\mathcal{V}$ and $\mathcal{U}$ are two orbifold atlases on a space M, that $\mathcal{V}$ is a refinement of $\mathcal{U}$ and that for every $V \in \mathcal{V}$ and $U \in \mathcal{U}$ with $V \subset U$ we have an injection $(\tilde{V}, \Delta, \phi) \hookrightarrow (\tilde{U}, \Gamma, \varphi)$, where $(\tilde{V}, \Delta, \phi)$ and $(\tilde{U}, \Gamma, \varphi)$ are the respective uniformizing charts. We then say that $\mathcal{V}$ and $\mathcal{U}$ are **directly equivalent orbifold atlases**. Now take the smallest equivalence relation on the orbifold atlases on M

---

[12]Ruan calls it a *uniformizing system*.

so that any two directly equivalent atlases on M are equivalent. We now define an **orbifold** to be a Hausdorff topological space together with an equivalence class of orbifold atlases.

Let x be a point in an orbifold M, and let $(\tilde{U}, \Gamma, \varphi)$ be a uniformizing chart with $x \in \varphi(\tilde{U})$. The **(orbifold) structure group** of x is the isotropy group of a point in the fiber $\varphi^{-1}(x)$. It is well-defined as an abstract group: if $x_1, x_2 \in \varphi^{-1}(x)$ then the corresponding isotropy groups are conjugate in Γ.

Remark III.39. It is not hard to show that if $(\tilde{U}, \Gamma, \varphi)$ is a uniformizing chart of U, then for any point $x \in U$ there is a neighborhood U' and a uniformizing chart $(\tilde{U}', \Gamma', \varphi')$ induced from $(\tilde{U}, \Gamma, \varphi)$ such that $(\varphi')^{-1}(x)$ is a single point $\tilde{x}$ (and hence Γ' fixes $\tilde{x}$). We will refer to $(\tilde{U}', \Gamma', \varphi')$ as a chart **centered** at x.

Let M be an orbifold with an atlas $\{\tilde{U}_i, \Gamma_i, \varphi_i\}$. A **smooth function** f on M is a collection of smooth Γ_i-invariant functions $\tilde{f}_i$ on $\tilde{U}_i$ such that for any injection $(\psi_{ij}, \lambda_{ij}) : (\tilde{U}_j, \Gamma_j, \varphi_j) \hookrightarrow (\tilde{U}_i, \Gamma_i, \varphi_i)$ we have $\psi_{ij}^* \tilde{f}_i = \tilde{f}_j$. Naturally each $\tilde{f}_i$ defines a continuous map $f_i : \tilde{U}_i/\Gamma_i = U_i \to \mathbb{R}$. Thanks to the compatibility conditions above, these maps glue together to define a continuous map from the topological space underlying M to $\mathbb{R}$. By abuse of notation we may write $f : M \to \mathbb{R}$. Similarly, a **differential k-form** σ on the orbifold M is a collection of Γ_i-invariant k-forms $\tilde{\sigma}_i$ on $\tilde{U}_i$ such that for any injection $(\psi_{ij}, \lambda_{ij}) : (\tilde{U}_j, \Gamma_j, \varphi_j) \hookrightarrow (\tilde{U}_i, \Gamma_i, \varphi_i)$ we have $\psi_{ij}^* \tilde{\sigma}_i = \tilde{\sigma}_j$. We denote the collection of all differential forms on M by $\Omega^*(M)$. Note that $\Omega^*(M)$ has a well-defined exterior multiplication (since exterior multiplication behaves well under pull-backs) and exterior differentiation $d : \Omega^*(M) \to \Omega^{*+1}(M)$ (for the same reason). In the same manner one defines vector fields, Riemannian metrics, quadratic forms and other differential geometric objects on orbifolds.

Finally we define maps of orbifolds. The reader should be aware that there are several notions of maps of orbifolds. For example Satake in his two papers gave two inequivalent definitions. The one we give below is the simplest; it is not the best. It will, however, suffice for our purposes. See Ruan's survey [R] for a more modern point of view. Let $(M, \{\tilde{U}_i, \Gamma_i, \varphi_i\})$ and $(N, \{\tilde{V}_j, \Delta_j, \phi_j\})$ be two orbifolds and let $F : M \to N$ be a continuous map of underlying topological spaces. The map F is a (smooth) **map of orbifolds** if for every point $x \in M$ there are charts $(\tilde{U}_i, \Gamma_i, \varphi_i)$, $(\tilde{V}_j, \Delta_j, \phi_j)$ with $x \in \varphi(\tilde{U}_i)$, $F(\varphi_i(\tilde{U}_i)) \subset \phi_j(\tilde{V}_j)$ and a C^∞ map $\tilde{F}_{ji} : \tilde{U}_i \to \tilde{V}_j$ such that

$$\phi_j \circ \tilde{F}_{ji} = F \circ \varphi_i.$$

The reason for the appearance of orbifolds in these notes is the symplectic reduction theorem of Marsden, Weinstein and Meyer [MW, Me]:

Theorem III.40. *Let (M, ω) be a symplectic manifold with a Hamiltonian action of a Lie group G. Let $\Phi : M \to \mathfrak{g}^*$ denote a corresponding moment map. Suppose $\eta \in \mathfrak{g}^*$ is a regular value of Φ and suppose that the action of the isotropy group G_η of η on $\Phi^{-1}(\eta)$ is proper. Then the action of G_η on $\Phi^{-1}(\eta)$ is locally free and the quotient $M_\eta := \Phi^{-1}(\eta)/G_\eta$ is naturally a symplectic orbifold.*

We will not discuss the proof of this well known theorem. Instead let me briefly explain why quotients by locally free actions are orbifolds. The reason is the slice theorem: if the action of a Lie group G is locally free and proper on a manifold Z, then for any point $z \in Z$ there is a slice S_z for the action of G and the isotropy group G_z is finite. The quotient S_z/G_z is homeomorphic to a neighborhood of the orbit $G \cdot z$ in the orbit space Z/G. The triple $(S_z, G_z, \varphi : S_z \to S_z/G_z \hookrightarrow Z/G)$ is a uniformizing chart of the orbifold Z/G.

Let us now prove Theorem III.38.

Proof of Theorem III.38. Since B is compact, the image $\Psi_\alpha(B)$ is compact. Therefore the set of vectors $X' \in \mathfrak{g}$, such that the function $\langle \Psi_\alpha, X' \rangle$ is strictly positive on B, is open. Hence we may assume that X lies in the integral lattice $\mathbb{Z}_G := \ker(\exp : \mathfrak{g} \to G)$ of the torus G. Let $H = \{\exp tX \mid t \in \mathbb{R}\}$ be the corresponding circle subgroup of G.

Let $f(x) = 1/(\langle \Psi_\alpha(x), X \rangle)$ and let $\alpha' = f\alpha$. The form α' is another G-invariant contact form with $\ker \alpha' = \xi$. The moment map $\Psi_{\alpha'}$ defined by α' satisfies $\Psi_{\alpha'} = f\Psi_\alpha$. Therefore $\langle \Psi_{\alpha'}(x), X \rangle = 1$ for all $x \in B$.

Since the function $\langle \Psi_\alpha, X \rangle$ is nowhere zero, the action of H on B is locally free. Consequently the induced action of H on the symplectization $(N, \omega) = (B \times \mathbb{R}, d(e^t \alpha'))$ is locally free as well. Hence any $a \in \mathbb{R}$ is a regular value of the X-component $\langle \Phi, X \rangle$ of the moment map Φ for the action of G on the symplectization (N, ω). Note that $\Phi(x, t) = -e^t \Psi_{\alpha'}(x)$. The manifold $B \times \{0\}$ is the -1 level set of $\langle \Phi, X \rangle$. Therefore by the symplectic reduction theorem $M := (\langle \Phi, X \rangle)^{-1}(-1)/H \simeq B/H$ is a (compact connected) symplectic orbifold. The action of G on $(\langle \Phi, X \rangle)^{-1}(-1)$ descends to an effective Hamiltonian action of G/H on M. A dimension count shows that the effective action of G/H on M is completely integrable, i.e., M is a symplectic toric orbifold. $\qquad\square$

The rest of the section is a proof that real odd dimensional cohomology vanishes for compact symplectic toric orbifolds. It uses Morse theory.

III.4.1   Morse theory on orbifolds

Let us start by briefly reviewing the fundamental results of Morse theory on manifolds. Let $f : M \to \mathbb{R}$ be a smooth function. A **critical point** of f is a point p where the differential df is 0. The image $f(p)$ of a critical point p is a **critical value** of f. A critical point p of f is **nondegenerate** if the Hessian $d^2 f_p$ is a nondegenerate quadratic form (in local coordinates $d^2 f_p$ is the matrix of second order partials $\left(\frac{\partial^2 f}{\partial x_i \partial x_j} \right)$; thus $d^2 f_p(x) = \sum_{ij} \frac{\partial^2 f}{\partial x_i \partial x_j} x_i x_j$). The **index** of a nondegenerate critical point p is the number of negative eigenvalues of the Hessian $d^2 f_p$ (counted with multiplicities.) The two and a half fundamental results of Morse theory are the two theorems and the lemma below.

Theorem III.41. *Let f be a smooth function on a manifold M, and M_a the set $f^{-1}((-\infty, a])$. If $f^{-1}([a, b])$ is compact and contains no critical points then M_a is homotopy equivalent to M_b.*

Sketch of proof. Fix a Riemannian metric on M. Since f has no critical points in $f^{-1}([a, b])$, the unit vector field $X = -\nabla f/\|\nabla f\|$ is well-defined. Extend X to all of M. The flow of X gives a retraction of M_b onto M_a. □

Theorem III.42. *Let f be a smooth function on a manifold M. Suppose $f^{-1}([a, b])$ is compact and contains exactly one nondegenerate critical point p in its interior. Then M_b has the homotopy type of M_a with a λ-dimensional disk D^λ attached along the boundary $\partial D^\lambda = S^{\lambda-1}$ where λ is the index of p.*

We omit the proof of the theorem which is well known noting only that the key ingredient of the proof is

Lemma III.43 (Morse Lemma). *Let p be a nondegenerate critical point of index λ of a function $f : M \to \mathbb{R}$. There is a neighborhood U of p in M and an open embedding $\varphi : U \to M$ such that $f \circ \varphi(x) = f(p) + d^2 f_p(x)$ for all $x \in U$. There is, there is a change of coordinates near p so that in new coordinates f is a quadratic form (up to a constant).*

Moreover, if a compact Lie group G acts on M fixing p and preserving f, we may arrange for U to be G-invariant and for φ to be G-equivariant. Note that in this case the Hessian $d^2 f_p(x)$ is a G-invariant quadratic form.

Suppose now that f is a smooth function on an *orbifold* M. Then the 1-form df still makes sense. We define p to be a **critical point** of f if $df_p = 0$. A critical point p is **nondegenerate** if for any uniformizing chart $(\tilde{U}, \Gamma, \varphi)$ with $p \in U = \varphi(\tilde{U})$, a point $\tilde{p} \in \varphi^{-1}(p)$ is a nondegenerate critical point of $\tilde{f} = f \circ \varphi$. The **index** of p is the index of the Γ-invariant quadratic form $d^2 \tilde{f}_{\tilde{p}}$.

It is not hard to believe that Theorem III.41 holds for orbifolds with no changes and that essentially the same proof still works. Theorem III.42 requires a small modification, see [LT].

Theorem III.44. *Let f be a smooth function on an orbifold M. Suppose $f^{-1}([a, b])$ is compact and contains exactly one nondegenerate critical point p in its interior. Then M_b has the homotopy type of M_a with the quotient D^λ/Γ attached along the boundary $(\partial D^\lambda)/\Gamma = S^{\lambda-1}/\Gamma$ where λ is the index of p and Γ is structure group of p. Here again D^λ and $S^{\lambda-1}$ denote the disk of dimension λ and the sphere of dimension $\lambda - 1$ respectively.*

Corollary III.45. *Let $f : M \to \mathbb{R}$, p, λ and Γ be as above. Then $H^*(M_b, M_a; \mathbb{R}) = H^*(D^\lambda/\Gamma, S^{\lambda-1}/\Gamma; \mathbb{R})$. Hence, if Γ is orientation preserving, $H^q(M_b, M_a; \mathbb{R}) = \mathbb{R}$ for $q = \lambda$ and 0 otherwise.*

Proof. By excision $H^*(M_b, M_a; \mathbb{R}) = H^*(D^\lambda/\Gamma, S^{\lambda-1}/\Gamma; \mathbb{R})$. If Γ is orientation preserving, then $H^q(D^\lambda/\Gamma, S^{\lambda-1}/\Gamma; \mathbb{R}) = \tilde{H}^q(S^\lambda/\Gamma; \mathbb{R}) = \mathbb{R}$ for $q = \lambda$ and 0 otherwise. Here $\tilde{H}^q$ denotes the reduced cohomology. □

As a consequence of the corollary above we get

Corollary III.46. *Let f be a smooth function on a compact orbifold M. Suppose all indices of f are even. Then $H^q(M, \mathbb{R}) = 0$ for all odd indices q.*

Proof. The proof is inductive. Suppose that c is a critical value of f and suppose we know that for all $a < c$ we have $H^q(M_a, \mathbb{R}) = 0$ for q odd. Assume for simplicity that $f^{-1}(c)$ contains only one critical point (this keeps the notation more manageable) and that its index is $2k$ for some integer k. Then using the long exact sequence for the pair $(M_{c+\epsilon}, M_{c-\epsilon})$ (for some sufficiently small $\epsilon > 0$) and Corollary III.45 we see that $H^q(M_{c+\epsilon}, \mathbb{R}) = H^q(M_{c-\epsilon}, \mathbb{R})$ for $q \neq 2k - 1, 2k$, that $H^{2k-1}(M_{c+\epsilon}, \mathbb{R})$ embeds in $H^{2k-1}(M_{c-\epsilon}, \mathbb{R}) = 0$ and that $H^{2k}(M_{c+\epsilon}, \mathbb{R}) = H^{2k}(M_{c-\epsilon}, \mathbb{R}) \oplus \mathbb{R}$. The result follows.   □

Now suppose $(M, \omega, \Phi : M \to \mathfrak{g}^*)$ is a compact symplectic toric orbifold. Just as for symplectic toric manifolds the image $\Phi(M)$ is a simple polytope and the moment map sends the fixed points M^G in one-to-one fashion to vertices of $\Phi(M)$ (see [LT]). Therefore, for a generic vector $X \in \mathfrak{g}$ the function $f = \langle \Phi, X \rangle$ takes distinct values at fixed points. We will now argue that the critical points of f are exactly the fixed points. We will then argue that f is Morse and that all indices of f are even.

For an action of a torus G on a compact orbifold M only finitely many isotropy groups can occur. This is a consequence of compactness and existence of slices. Therefore the set of subalgebras $\mathfrak{g}_x$, which are Lie algebras of isotropy groups for the action of G on M, is finite. Hence the set

$$\mathcal{U} = \mathfrak{g} \setminus \bigcup_{x \in M} \{\mathfrak{g}_x \mid \mathfrak{g}_x \neq \mathfrak{g}\}$$

is open and dense. Now take X to be a vector in $\mathcal{U}$. Then

$$0 = df_x = d\langle \Phi, X \rangle_x = (\iota(X_M)\omega)_x \Leftrightarrow X_M(x) = 0$$
$$\Leftrightarrow \exp tX \cdot x = x \text{ for all } t$$
$$\Leftrightarrow \exp tX \in G_x \text{ for all } t$$
$$\Leftrightarrow X \in \mathfrak{g}_x.$$

Since $X \in \mathcal{U}$ we see that $df_x = 0 \Leftrightarrow \mathfrak{g}_x = \mathfrak{g} \Leftrightarrow x$ is fixed by G. It remains to check that f is Morse and that all the indices of f are even.

Let $x \in M^G$ be a fixed point and let $(\tilde{U}, \Gamma, \varphi)$ be a uniformizing chart centered at x (cf. Remark III.39). We may assume that $U = \varphi(\tilde{U})$ is G-invariant. Denote the symplectic form on $\tilde{U}$ by $\tilde{\omega}$. The map $\tilde{\Phi} = \Phi \circ \varphi : \tilde{U} \to \mathfrak{g}^*$ is a moment map for an action of a torus $\tilde{G}$ on $\tilde{U}$. One can show arguing as in the proof of Lemma III.6 that $\Gamma \subset \tilde{G}$ and that $G = \tilde{G}/\Gamma$. In particular $\tilde{G}$ has $\mathfrak{g}$ as its Lie algebra. We now apply the equivariant Darboux theorem to $\tilde{U}$, $\tilde{\omega}$ and $\tilde{G}$. We get a $\tilde{G}$ invariant neighborhood $\tilde{V}$ of $\tilde{x} = \varphi^{-1}(x)$, an open neighborhood $\tilde{V}_0$ of 0 in

$T_{\tilde{x}}\tilde{U}$ and a $\tilde{G}$-equivariant diffeomorphism $\tau : \tilde{V}_0 \to \tilde{V}$ such that $\tau^*\tilde{\omega} = \tilde{\omega}_0$ where $\tilde{\omega}_0$ is the constant coefficient form $\tilde{\omega}_{\tilde{x}}$ on the vector space $T_{\tilde{x}}\tilde{U}$.

The action of $\tilde{G}$ on $T_{\tilde{x}}\tilde{U}$ is linear. Hence the corresponding moment map $\tilde{\Phi}_0 : T_{\tilde{x}}\tilde{U} \to \mathfrak{g}^*$ is quadratic. In fact by choosing a $\tilde{G}$-invariant complex structure on $T_{\tilde{x}}\tilde{U}$ compatible with the symplectic form we can identify $(T_{\tilde{x}}\tilde{U}, \tilde{\omega}_0)$ with $(\mathbb{C}^n, \sqrt{-1}\sum dz_j \wedge d\bar{z}_j)$ so that the action of $\tilde{G}$ is given by

$$a \cdot (z_1, \ldots, z_n) = (\chi_1(a)z_1, \ldots, \chi_n(a)z_n)$$

for some characters $\chi_j : \tilde{G} \to S^1$. Then

$$\tilde{\Phi}_0(z_1, \ldots, z_n) = \sum |z_j|^2 \nu_j$$

where $\nu_j = d\chi_j$ are the corresponding weights. Since τ is a $\tilde{G}$-equivariant symplectomorphism, $\tilde{\Phi} \circ \tau = \tilde{\Phi}_0 + c$ for some constant $c \in \mathfrak{g}^*$. Hence

$$f \circ \varphi \circ \tau = \langle \Phi, X \rangle \circ \varphi \circ \tau = \langle \tilde{\Phi}, X \rangle \circ \tau = \langle \tilde{\Phi}_0, X \rangle + c, \text{ i.e.,}$$

$$f \circ \varphi \circ \tau(z_1, \ldots, z_n) = \sum \nu_j(X)|z_j|^2 + c.$$

By the choice of X, $\nu_j(X) \neq 0$ for any j. Thus $f \circ \varphi$ is Morse and its index at $\tilde{x}$ is twice the number of weights ν_j with $\nu_j(X) < 0$. Therefore, by definition, f is a Morse function on M and all indices of f are even.

Combining the above discussion with Corollary III.46 we get

Proposition III.47. *Let M be a compact symplectic toric orbifold. Then for any odd index q the cohomology group $H^q(M, \mathbb{R})$ is zero.*

Appendix

Hypersurfaces of contact type

Contact manifolds often arise as codimension 1 submanifolds of symplectic manifolds, i.e., as hypersurfaces.

Definition III.48. Let (M, ω) be a symplectic manifold. A hypersurface Σ of M is of **contact type** if there is a neighborhood U of Σ in M and a vector field X on U such that

1. $T_m M = T_m \Sigma \oplus \mathbb{R} X(m)$ for any point $m \in \Sigma$, i.e., the vector field X is nowhere tangent to Σ;

2. the flow of X expands the symplectic form ω exponentially, i.e., $L_X \omega = \omega$, where as usual L_X denotes the Lie derivative with respect to X.

The vector field X with above properties is often called a **Liouville vector field**.

We now prove that hypersurfaces of contact type are indeed contact manifolds.

Proposition III.49. *Let Σ be a hypersurface of contact type in a symplectic manifold (M, ω) and let X be a Liouville vector field defined on a neighborhood U of Σ. The 1-form $\alpha = (\iota(X)\omega)|_\Sigma$ is contact.*

Proof. Note first that $d(\iota(X)\omega) = d(\iota(X)\omega) + \iota(X)d\omega = L_X \omega = \omega$. Hence $d\alpha = d(\iota(X)\omega)|_\Sigma = \omega|_\Sigma$. Since ω is symplectic and Σ is of codimension 1 in M, the form $\omega_m|_{T_m \Sigma}$ has a 1-dimensional kernel (for any point $m \in \Sigma$). Thus there is a vector $Y_m \in T_m \Sigma$ such that $\omega_m(Y_m, v) = 0$ for any $v \in T_m \Sigma$. Since ω is symplectic and since by assumption $T_m M = T_m \Sigma \oplus \mathbb{R} X(m)$ for all $m \in \Sigma$ we have

$$0 \neq \omega_m(Y_m, X(m)) = -(\iota(X)\omega)_m(Y_m) = -\alpha_m(Y_m).$$

Hence $\alpha_m \neq 0$ for all $m \in \Sigma$ and consequently $\xi = \ker \alpha$ is a codimension 1 distribution on Σ. It remains to show that for any $m \in \Sigma$

$$d\alpha_m|_{\xi_m} = \omega_m|_{\xi_m}$$

is non-degenerate. On the other hand $d\alpha_m = \omega_m|_{T_m \Sigma}$ and $\xi_m = \{v \in T_m \Sigma \mid \omega_m(v, X(m)) = 0\}$. Since $\omega_m(v, Y_m) = 0$ for any $v \in T_m \Sigma$, the subspace ξ_m of $T_m M$ lies in the symplectic perpendicular to the 2-plane $\mathrm{span}_\mathbb{R}\{Y_m, X(m)\}$ in $(T_m M, \omega_m)$:

$$\xi_m \subset (\mathrm{span}_\mathbb{R}\{Y_m, X(m)\})^\omega.$$

Since $\dim \xi_m = \dim M - 2 = \dim(\mathrm{span}_\mathbb{R}\{Y_m, X(m)\})^\omega$, we have equality: $\xi_m = (\mathrm{span}_\mathbb{R}\{Y_m, X(m)\})^\omega$. Since $(\mathrm{span}_\mathbb{R}\{Y_m, X(m)\})$ is a symplectic subspace, its symplectic perpendicular ξ_m is symplectic as well. $\qquad\square$

Example III.50. Let (Q, g) be a Riemannian manifold. Let g^* denote the dual metric on T^*Q. The co-sphere bundle S^*Q is the set of covectors in T^*Q of length 1: $S^*Q = \{(q, p) \in T^*Q \mid q \in Q, p \in T_q^*Q, g_q^*(p, p) = 1\}$. It is a hypersurface of contact type in T^*Q relative to the standard symplectic structure. The Liouville vector field is the generator of dilations. In local coordinates $X(q, p) = \sum p_j \frac{\partial}{\partial q_j}$.

Exercise III.51. A codimension 1 hypersurface $\Sigma \subset \mathbb{C}^n$ is **star-shaped about the origin** if for any nonzero vector $v \in \mathbb{C}^n$ the ray $\{tv \mid t \in (0, \infty)\}$ intersects Σ transversely in exactly one point. In particular Σ is the image of an embedding $\iota : S^{2n-1} \hookrightarrow \mathbb{C}^n$ of the $(2n - 1)$-dimensional sphere.

Show that any star-shaped hypersurface is a hypersurface of contact type in $(\mathbb{C}^n, \omega = \sqrt{-1} \sum dz_j \wedge d\bar{z}_j)$. Show that any two star-shaped hypersurfaces are isomorphic as contact manifolds. The contact structure in question is called the **standard contact structure on** S^{2n-1}.

Prove a converse: given a contact form α on S^{2n-1} defining the standard contact structure there is an embedding $\iota : S^{2n-1} \to \mathbb{C}^n$ such that $\iota^*(\iota(X)\omega) = \alpha$ where X is the radial vector field on $\mathbb{C}^n$: $X(z) = \frac{1}{2} \sum \left(z_j \frac{\partial}{\partial z_j} + \bar{z}_j \frac{\partial}{\partial z_j} \right)$.

Bibliography

[A]   M. F. Atiyah, Convexity and commuting hamiltonians, *Bull. London Math. Soc.* **14** (1982), 1-15.

[B]   A. Banyaga, The geometry surrounding the Arnold-Liouville theorem in *Advances in geometry*, edited by Jean-Luc Brylinski, Ranee Brylinski, Victor Nistor, Boris Tsygan and Ping Xu. Progress in Mathematics, 172. Birkhäuser Boston, Inc., Boston, MA, 1999. xii+399 pp. ISBN 0-8176-4044-4

[BM1]   A. Banyaga and P. Molino, Géométrie des formes de contact complètement intégrables de type toriques. Séminaire Gaston Darboux de Géométrie et Topologie Différentielle, 1991 1992 (Montpellier), 1–25, Univ. Montpellier II, Montpellier, 1993.

[BM2]   A. Banyaga and P. Molino, Complete integrability in contact geometry, Penn State preprint PM 197, 1996.

[BoM]   M. Boucetta and P. Molino, Géomtrie globale des systèmes hamiltoniens complètement intégrables: fibrations lagrangiennes singulières et coordonnées action-angle à singularités, *C. R. Acad. Sci. Paris Sér. I Math.* **308** (1989), no. 13, 421–424.

[BG]   C. P. Boyer and K. Galicki, A note on toric contact geometry, *J. Geom. Phys.* **35** (2000) 288–298; math.DG/9907043v2.

[BI]   D. Burago and S. Ivanov, Riemannian tori without conjugate points are flat, *GAFA*, 4:3 (1994), 259–269.

[D]   T. Delzant, Hamiltoniens périodiques et images convexes de l'application moment, *Bull. Soc. Math. France* **116** (1988), no. 3, 315–339.

[Ge]   H. Geiges, Constructions of contact manifolds, *Math. Proc. Cambridge Philos. Soc.* **121** (1997), no. 3, 455–464.

[GS1]   V. Guillemin and S. Sternberg, Convexity properties of the moment mapping I, *Invent. Math.* **67** (1982), 491-513.

[GS]   V. Guillemin and S. Sternberg, *Symplectic techniques in physics*, Cambridge University Press, Cambridge–New York, 1984. xi+468 pp. ISBN: 0-521-24866-3

[HS]   A. Haefliger and E. Salem, Action of tori on orbifolds, *Ann. Global Anal. Geom.* **9** (1991), 37–59.

[L1]   E. Lerman, Contact cuts, *Israel J. Math* **124** (2001), 77–92. See also http://xxx.lanl.gov/abs/math.SG/0002041.

[L2]   E. Lerman, A convexity theorem for torus actions on contact manifolds, *Illinois J. Math.*, to appear. See http://xxx.lanl.gov/abs/math.SG/0012017.

[LS]   E. Lerman and N. Shirokova, Completely integrable torus actions on symplectic cones, *Math. Res. Lett.* **9** (2002), no. 1, 105–115. See also http://xxx.lanl.gov/abs/math.DG/0011139.

[LT]   E. Lerman and S. Tolman, Symplectic toric orbifolds, *Trans. A.M.S.* **349** (1997), 4201–4230.

[M]   R. Mañe, On a theorem of Klingenberg, in *Dynamical systems and bifurcation theory*, 319–345, Pitman Res. Notes Math. Ser., **160**, Longman Sci. Tech., Harlow, 1987.

[MW]   J. Marsden and A. Weinstein, Reduction of symplectic manifolds with symmetry, *Rep. Math. Phys.* **5** (1974), 121–130.

[Me]   K. Meyer, Symmetries and integrals in mechanics, *Dynamical systems* (Proc. Symp., Univ. Bahia, Salvador, 1971), 259–272. Academic Press, New York, 1973.

[P]   R. Palais, The classification of *G*-spaces, *Mem. Amer. Math. Soc.* **36** (1960).

[R]   Y. Ruan, Stringy geometry and topology of orbifolds, http://xxx.lanl.gov/abs/math.AG/0011149

[Sa1]   I. Satake, On a generalization of the notion of manifold, *Proc. Nat. Acad. Sc. USA* **42** (1956), 359–363.

[Sa2]   I. Satake, The Gauss-Bonnet theorem for V-manifolds *J. Math. Soc. Japan* **9** (1957), 464–492.

[TZ]   J. Toth and S. Zelditch, Riemannian manifolds with uniformly bounded eigenfunctions, *Duke Math Journal* **111** (2002), no. 1, 97–132. See also `http://xxx.lanl.gov/abs/math-ph/0002038`.

[WW]   R.S. Ward and R.O. Wells jr., *Twister Geometry and Field Theory*, Cambridge Univ. Press, Cambridge, 1990.

Advanced Courses in Mathematics CRM Barcelona

Since 1995 the Centre de Recerca Matemàtica (CRM) in Barcelona has conducted a number of annual Summer Schools at the post-doctoral or advanced graduate level. Sponsored mainly by the European Community, these Advanced Courses have usually been held at the CRM in Bellaterra.
The books in this series consist essentially of the expanded and embellished material presented by the authors in their lectures.

For orders originating from all over the world except USA and Canada:
Birkhäuser Verlag AG
c/o Springer GmbH & Co
Haberstrasse 7
D-69126 Heidelberg
Tel.: ++49 / 6221 / 345 0
Fax: ++49 / 6221 / 345 4229
e-mail: birkhauser@springer.de

For orders originating in the USA and Canada:
Birkhäuser
333 Meadowland Parkway
USA-Secaucus, NJ 07094-2491
Fax: +1 201 348 4033
e-mail: orders@birkhauser.com

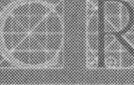

CENTRE DE RECERCA MATEMÀTICA

Birkhäuser   http://www.birkhauser.ch